Responsible Brains

Responsible Brains

Neuroscience, Law, and Human Culpability

William Hirstein, Katrina L. Sifferd, and Tyler K. Fagan

The MIT Press
Cambridge, Massachusetts
London, England

This book was set in ITC Stone Serif Std by Toppan Best-set Premedia Limited. Printed and bound in the United States of America.

Library of Congress Cataloging-in-Publication Data

Names: Hirstein, William, author. | Sifferd, Katrina, author. | Fagan, Tyler, author.
Title: Responsible brains : neuroscience, law, and human culpability / William Hirstein, Katrina L. Sifferd, and Tyler K. Fagan.
Description: Cambridge, MA : MIT Press, [2018] | Includes bibliographical references and index.
Identifiers: LCCN 2018008742 | ISBN 9780262038782 (hardcover : alk. paper)
Subjects: LCSH: Responsibility. | Brain.
Classification: LCC BJ1451 .H57 2018 | DDC 153--dc23 LC record available at https://lccn.loc.gov/2018008742

10 9 8 7 6 5 4 3 2 1

Contents

Preface

When we praise, blame, reward, or punish a person for doing something, it is usually because we think that person is *responsible* for that action—and hence that the person is, in some significant sense, the author of his or her actions and thus deserves to be held accountable for them, as well as their consequences. Making these assignments of responsibility is utterly central to human life, yet exactly how to justify them remains an open question.

Common sense tells us that what makes human beings responsible has something to do with their minds, and with the relations between their minds and their actions. And we now have good reason to think that the brain accomplishes the functions of the mind, thanks to an era of rapid progress in neuroscience which began in the 1980s and has continued to gather momentum. This means that the issues surrounding responsibility can now be approached from a relatively secure empirical standpoint, from a branch of biology—neurobiology. Given this, if a defendant, on trial for a horrible murder, were found to have serious brain damage, which brain parts or processes would have to be damaged for this defendant to be considered less responsible, or not responsible at all? Why do certain mental illnesses, with their corresponding profiles of brain dysfunction, seem to justify pleas of legal excuse? What is it about the developing brains of children and adolescents that makes them less culpable than adults, if not excused from responsibility altogether?

If a person's mind is damaged or still developing, or if the person's actions are fundamentally uncoupled from the person's mental processes, we may judge the person less responsible for what he or she does, or perhaps excuse the person from responsibility altogether. Indeed, the idea of a "guilty mind"—mens rea—lies at the core of the criminal law, a stable and codified system of practices about whom to hold responsible, and for what. It seems obvious that the ability to reason has a great deal to do with being responsible, but exactly what sort of thinking or reasoning abilities are relevant in this way, and which brain processes accomplish them? What other brain functions are required for responsibility? Do all of these functions together constitute a meaningful functional unit in the brain? Could one actually point to a place in the brain (or a network of such places) as the seat of human responsibility?

We believe that the answer to these last two questions is yes. In an attempt to answer the question of where the mechanisms of responsibility reside in the human brain, we have constructed a comprehensive hypothesis about human responsibility couched ultimately in the language of neuroscience. In this book, we describe this theory, then put it to work addressing the above questions, comparing it with competing theories at each point.

This book is written to be accessible to virtually anyone interested in the above questions. Lawyers and other legal professionals should find our commonsense approach to responsibility congenial and practical. Medical professionals who make determinations of competency may also find this book useful. Within the academic community, the book should especially appeal to philosophers working on issues surrounding the concept of responsibility. More broadly, teachers or students in any of the cognitive sciences, including philosophy, psychology, linguistics, neuroscience, and artificial intelligence, who are interested in how their fields connect to issues in ethics may find this book worthwhile.

Acknowledgments

This book was made possible through the support of a grant from the John Templeton Foundation, although the opinions expressed here are our own and do not necessarily reflect the views of the John Templeton Foundation. We owe a great debt to Al Mele, the coordinator of the Philosophy and Science of Self-Control project, and manager of our sub-grant. His project not only funded our research for this book, but also provided venues at which our work could be presented.

We are also grateful to the many philosophers and scientists whose valuable input and feedback helped to shape this book—especially Neil Levy, who provided detailed comments on the chapters that analyze his theory of responsibility. We are deeply thankful to Philip Laughlin at the MIT Press for his guidance and open-mindedness during the book proposal process. We thank Elmhurst College for its support during this project, and we especially thank the students who participated in the *Responsible Brains* research group at Elmhurst College, including Daniel Hayes, Elana Hunt, Sean McKay, Kit Rempala, Margaret Sumney, Brandon Thompson, and Michael White.

Finally, we are grateful to our families for their love, support, and patience—with special thanks to Elsa, Harrison, Joanna, and Sally.

1 Introduction

Case Descriptions

Dominic Ongwen

It is the year 1990. A skinny boy named Dominic, perhaps ten or twelve years old, is abducted on his way to school in the Gulu District of northern Uganda. The men who have taken him are soldiers in the Lord's Resistance Army (LRA), a cultlike guerilla force led by the self-proclaimed prophet Joseph Kony. Dominic's abductors literally carry him to the LRA base; he is too small to keep up with them on foot.[1]

The son of schoolteachers, Dominic is now made to live with a middle-aged LRA commander and to think of this man as a military superior, mentor, and surrogate father. He is told to forget his parents and former life, and taught to empty his mind so it can be filled with passionate devotion to the LRA's holy cause. His indoctrination includes disorienting, exhausting periods of marching and hard labor punctuated by rituals of spiritual purification. In his daily intake of political and religious propaganda, he is told of Kony's supernatural powers, including the ability to detect any sign of disloyalty, or even private misgivings. Punishment is harsh and capricious, perversely fostering blind trust in the group's leaders and paranoia about nearly everyone else. Dominic witnesses murder, enslavement, and rape used as tools of war and discipline and is himself subjected to regular beatings, sometimes for minor infractions and sometimes for no reason at all. He sees other recruits and captives beaten, mutilated, and killed.

Over the next few years, Dominic's part in this milieu of violence shifts gradually from that of a passive witness to an eager accomplice and then to a primary enforcer. His military training intensifies, and he quickly shows himself to be a natural warrior—brave, intelligent, keen both to please his superiors and to assume his own command. According to charges filed in the International Criminal Court, he begins to plan and execute his own raids and abductions and engages in the same forms of indoctrination used on his younger self. The LRA measures a soldier's loyalty by how many people he has killed or kidnapped, and in this Dominic proves exceptional. He is also lucky: While many of his peers and superior officers are killed, Dominic survives to adulthood. By 18 he is a field commander, and by 25 he has secured a senior rank in the LRA's high command, his promotion allegedly buttressed by an unknown number of brutal attacks, massacres, and abductions.

In 2005, the International Criminal Court (ICC) charged Dominic Ongwen with crimes against humanity, including the murder and enslavement of civilians. He surrendered to authorities in 2015 and is, as of this writing, on trial at the ICC. In the time between the first charges and his surrender, prosecutors have expanded the list of his charges to include over 70 criminal counts. Because he is known to have suffered many of the very crimes he is now accused of committing, Ongwen's case is unusual—he is the first figure of his kind to be subject to international criminal prosecution—but he is not unique in illustrating difficult questions about the culpability of child soldiers, many of whom appear to be both victims and perpetrators.[2]

Reviewing the evidence against Ongwen, it seems likely he has been involved in dozens, perhaps hundreds of abductions and murders. But how many of these crimes is he truly responsible for? The brutalities enumerated in his arrest warrants occurred during his early 20s, although he had surely caused many deaths before that time—and it is hard to see any of those deeds as disconnected from Ongwen's childhood abduction and forcible recruitment into the LRA. Pinning down the precise details of Ongwen's biography has proven difficult, and it is unclear whether it matters exactly how old he was at the time of his kidnapping, or how old he was when he began to commit the crimes he is charged with. We know that many child soldiers, rather than participate actively in atrocities, try to "fade into the background," risk escape, or even choose death; does it matter that Ongwen seems instead to have embraced the goals and tactics of the LRA?

Anders Breivik

It is midday on July 22, 2011, and the final pieces of Anders Breivik's plan are clicking into place. By nightfall he will be in police custody, and 77 innocent persons will be dead by his hand.[3]

At around 2:00 in the afternoon, Breivik sends out a mass e-mail containing a manifesto that sprawls over 1,500 pages. In it, he advocates violence against Muslims and rails against the multiculturalism and "cultural Marxism" he sees as having infected European society. In the manifesto, and in subsequent interrogations, Breivik claims to be acting on behalf of a secret Christian military order, the Knights Templar, with dozens or hundreds of members throughout Western Europe. There is no evidence, however, that any such organization exists.

Shortly after sending the e-mail, the 32-year-old Breivik leaves his mother's flat in west Oslo, driving a white van into the government quarter. The van contains a bomb made of fertilizer and fuel oil, weighing over a ton, which he has spent the past months assembling. He is dressed in a homemade police uniform. He carries a pistol and a semiautomatic rifle, both acquired over years of planned, careful maneuvering through the legal conditions necessary for gun ownership in Norway. Breivik parks the van, lights a seven-minute fuse, and walks away. He gets into a different car, parked nearby the night before as part of his plot, and drives out of the city. At 3:25 p.m., the bomb explodes, killing eight people and injuring many more.

An hour and a half later, some 25 miles northwest of Oslo, Breivik boards a ferry to the island of Utøya, where roughly 600 youths are attending a summer camp organized by Norway's Labor Party. The island is isolated and tiny, only a little larger than Rockefeller Center in New York City. Minutes after arriving on Utøya, and appearing trustworthy in his false uniform, Breivik draws his weapons and begins killing people. Over the next hour he fires hundreds of rounds. He lines up campers against the wall of a cabin and executes them. He shoots teenagers as they attempt to swim away from the island. He discovers young men and women feigning death, by lying on the ground among the slain, and shoots them in the head. Survivors will later report that Breivik persuaded those hidden to come forward by saying he was a policeman, and that he was laughing and shouting as he killed them. After 50 minutes, Breivik calls the police saying: "Yes, hello, my name is Commander Anders Behring Breivik from the Norwegian anti-communist resistance movement. I'm on Utøya for the moment. I want to give myself up" (Melle 2013). After this call he continues shooting. In his first interrogation after being taken into custody, Breivik again claims to be commander of the Knights Templars Norway and states that "the people on the island were category C traitors" (Melle 2013). What happened on the island was unfortunate, he tells authorities, because he had the right to "kill category A and B traitors, but not a mandate to kill category C traitors" (Melle 2013). Between the Oslo bombing and the Utøya murders, Breivik kills 77 people that day, 55 of them teenagers.

Norway's legal system employs what is sometimes called the "medical model" of legal insanity: If a defendant is judged to have been psychotic at the time he or she committed a criminal act, then that defendant is to be excused from punishment (Moore 2014). Such judgments depend heavily, if not entirely, on psychiatric testimony. Breivik's initial review, conducted by a team of psychiatrists, found that he was psychotic—and thus legally insane—because he was suffering from paranoid schizophrenia. Many of Breivik's outlandish beliefs were construed, through the lens of this diagnosis, as "persistent, systematized, bizarre delusions" (Bortolotti, Broome, and Mameli 2014). But a second review, conducted by a different team of psychiatrists, found that Breivik suffered from antisocial and narcissistic personality disorders but *not* paranoid schizophrenia and was therefore not psychotic. The panel of judges hearing Breivik's case found this second review more persuasive, and Breivik was judged legally sane and sentenced to 21 years in prison, a term that can in principle be renewed until Breivik's death.

From the calculated horror of Breivik's crimes to the bizarre and delusional beliefs he expressed before and after the attacks, to the conflicting psychiatric diagnoses and Norwegian law's unusual deference to medical expertise in determining legal insanity—the facts of Breivik's case raise a number of troubling and thorny questions. We may find ourselves pulled in contrary directions, feeling on the one hand that any person who planned and executed these crimes deserves the harshest punishment possible under the law; and on the other hand that it would be unjust to punish someone who does not understand or cannot control his behavior and is therefore undeserving of blame. What was the state of Anders Breivik's sanity at the time of his crime, and how and why does the answer to that question matter? If Breivik was truly suffering from paranoid schizophrenia in 2011, does that mean he was not morally responsible for his actions and is therefore undeserving of legal punishment? What is the relevance of mental illness in general, or a specific mental illness in particular, to responsibility? How much evidentiary weight, if any, should a psychiatric diagnosis carry in deciding whether a defendant can properly be said to deserve punishment for his or her crime?

Brian Dugan

It is February 1983, and Brian Dugan is cruising aimlessly around the western Chicago suburbs.[4] Now 26 years old, he has spent the last ten years in and out of prison and the legal system, charged with an array of increasingly violent offenses, from burglary and battery to arson and attempted kidnapping. Jeanine Nicarico is alone at her house in Naperville, Illinois, home from school with the flu. She is ten years old. Dugan has been knocking on doors in the neighborhood, half-randomly, hoping to find an unoccupied house where he can break in and steal some tools. He knocks on Jeanine's door, and although she refuses to let him in, her way of answering Dugan's questions leaves no doubt that she is home alone. Dugan breaks in the front door, abducts Jeanine, rapes her, and beats her to death with a tire iron. Jeanine's body is found two days afterward. The shock of this heinous crime, committed in broad daylight on a quiet suburban street, convulses authorities into a frantic—and arguably incautious—search for the perpetrator. Pretending to have knowledge of the case in a misguided attempt to claim the $10,000 reward, a young man named Rolando Cruz winds up attracting police suspicion and finds himself arrested and charged with Jeanine's murder, along with a man named Alejandro Hernandez. With suspects in custody, Brian Dugan is free to resume his hunting.

In July of 1984, Dugan sideswipes a young woman's car, running her off the road. She is Donna Schnorr, a 27-year-old nurse, and a stranger to Dugan. He forces her into his car, binds her, and drives her to an abandoned quarry a few miles away, where he rapes and drowns her. Less than a year after that, in May of 1985, Dugan—not yet a suspect in the Schnorr killing—commits a string of intermittently successful abductions and rapes, the victims of which are all young women or girls. On June 2, Dugan's spree reaches its final phase; he kidnaps seven-year-old Melissa Ackerman, rapes her, and drowns her. A few weeks later her body is found, and Dugan is arrested shortly afterward; another young girl had been with Melissa at the time of the abduction, but had narrowly escaped, and she had been able to describe Dugan's car to police.

Under questioning, Dugan confesses to the Ackerman and Schnorr murders, each of which earns him a life sentence. He informally confesses to the Nicarico murder as well, but will only offer a formal confession if he can avoid the death penalty; prosecutors reject his offer. Eventually, DNA evidence is developed linking Dugan to Jeanine Nicarico's death; he is indicted for her murder in 2005 and pleads guilty in 2009, hoping to avoid the death penalty. As the sentencing phase of his trial begins, Dugan's attorneys are casting about for any evidence that might help mitigate their client's guilt in the eyes of the jury and keep him off death row. They learn of a cognitive neuroscientist in New Mexico named Kent Kiehl, who is an expert on psychopaths. Kiehl has spent years collecting data on the neuropsychology of violent offenders. He claims that the brains of psychopaths show distinctive patterns of dysfunction and decreased activity, particularly in areas responsible for integrating emotion with cognition, which explain why psychopaths often appear to be devoid of empathy, remorse, or sensitivity to the harms they cause. Dugan's lawyers reach out to Kiehl, offering him the chance to study, in Kiehl's words, "one of the classic psychopaths in American history" (Hughes 2010, 342).

Kiehl accepts the offer and sets to work, gaining access to Dugan's case files and conducting lengthy interviews with him. He scans Dugan's brain using functional magnetic resonance imaging (fMRI) and tests him on the Hare Psychopathy Checklist, an interview technique meant to detect the presence of psychopathy through 20 distinct personality and behavioral traits. Dugan scores a 38 out of 40 on the test, placing him, in Kiehl's estimation, above the 99th percentile of all prison inmates. Dugan's files reveal a childhood that bears the typical marks of a budding psychopath: cruelty to animals, chronic bed-wetting, and fascination with fire.

In October of 2009, Kiehl participates in a Frye hearing, a legal proceeding meant to assess whether some body of scientific evidence can be admitted at trial. The prosecution argues that allowing Kiehl to testify using the actual brain scans could unduly bias the jury, while the defense argues that the scans are necessary for a complete and accurate picture of Dugan's mental state. The judge finds a middle ground of sorts, allowing Kiehl to share his findings with the jury but not to use the actual brain scans obtained from Dugan. Kiehl takes the stand in November; his testimony is used to ground the defense's argument that Dugan, as a psychopath, was not fully in control of himself at the time of his crimes and was not capable of suppressing or managing his powerful impulses to harm, violate, and kill. The horrific rape and murder of Jeanine Nicarico was perpetrated not by an evil man, but by a sick man, they argue, and his condition calls for a more lenient sentence, or at least something short of execution. The prosecution counters Kiehl's testimony with their own expert witnesses, who point out the evidential limitations of fMRI and argue that a brain scan taken 26 years after the crime in question can hardly illuminate Dugan's mental state at the time of Nicarico's murder. Even if Kiehl had ironclad evidence of Dugan's psychopathy—not just now but back in 1983—there should be serious doubts about whether psychopathy should eo ipso mitigate a defendant's blameworthiness in the eyes of the law. The jury in Dugan's case returns the unanimous verdict needed for the death penalty, and Brian Dugan is sentenced to die.

In March 2011, however, Illinois Governor Pat Quinn signed a bill abolishing the state's use of the death penalty and commuting the death sentences of 15 Illinois prisoners—Dugan among them. The death penalty had already been under a statewide moratorium since 2000, and Quinn's action was the culmination of concerns about wrongful convictions and executions of the innocent. In a bizarre twist, two of the most notorious wrongful convictions were those of Rolando Cruz and Alejandro Hernandez, who had both been convicted and sentenced to death for the rape and murder of Jeanine Nicarico. Had Brian Dugan not made his informal confession back in 1985, the two men might well have been executed. (Gutowski and Mills 2014).

Today Brian Dugan is 61 years old, serving life without the possibility of parole in the Stateville Correctional Center, less than 20 miles from the scene of his first murder. He appears to have evaded the death penalty for good, but he will die in prison. In his first published interview, in 2014, Dugan told reporters from the *Chicago Tribune* that he still felt he was a danger to society, prone to intense and uncontrollable rage. Describing the murders, he said, "I was driven by some kind of an impulse that kept growing. I could not stop" (Gutowski and Mills 2014).

The case of Brian Dugan presents a number of difficult and unsettling questions: How should we understand the moral responsibility, or lack thereof, of the psychopath? Are psychopaths "more bad than mad," as Maibom (2008) argues, or is it the other way around? To what extent do psychopaths understand the wrongness or harmfulness of their actions? If they seem unmoved or unmotivated by moral norms, can they really be said to *understand* them? And what about volitional control, often described as a crucial component of moral responsibility—are psychopaths truly in control of their actions? What are the cognitive and neurological signatures of psychopathy? What sorts of evidence from psychiatry and neuroscience would indicate that a defendant was psychopathic, and should that same evidence tend to reduce, or excuse the defendant from, culpability? If there is such evidence to be had, how should it be responsibly introduced into the courtroom?

Responsible Brains

The case studies above involve people who caused serious harm, but to whom attributing responsibility for that harm seems particularly difficult. They are cases where pertinent-seeming facts about the agents' minds muddle the intuitions about agency and intent that underpin attributions of responsibility and, therefore, blame. As a result, courts have disagreed about the level of blame and punishment that ought to be assigned to these sorts of agents—for instance, child soldiers (or children, generally) and the mentally ill, including psychopaths.

In this book we will argue that specific facts about the brains of the agents discussed in these cases, now available due to recent progress in neuroscience, can strongly inform assessment of their culpability. This may not sound like a bold claim: It is not very controversial, after all, to say that facts about people's mental capacities and mental states may be relevant to their responsibility for harmful acts. And it is similarly commonplace to believe that facts about an agent's brain are relevant to facts about that agent's mind. Some scientists and philosophers would even prefer to say that facts about the mind simply are facts about the brain because mental states simply are a type of brain state.[5]

However, one needn't go this far to find persuasive the idea that learning about how the brain works can tell us about how the mind works, or that facts about a person's brain might bear on his or her responsibility. Agents who have a certain level of plaque and tangles in the brain—both of which are highly correlated with Alzheimer's disease—can be expected to suffer from certain mental incapacities. The link between brain plaque and the symptoms of Alzheimer's is strong enough for facts about levels of plaque and tangles to be relevant in at least certain cases of attributing responsibility. For example, imagine Bob is accused of the theft of a briefcase. Bob claims his Alzheimer's disease led him to forget that the briefcase was not his. However, the victim of the theft claims Bob took the briefcase intentionally and is malingering (is just pretending he has Alzheimer's). If new brain imaging techniques allowed a medical specialist to provide good evidence of plaque buildup or tangles in Bob's brain, this would obviously be relevant to his excuse.

Evidence of traumatic brain injuries or other indications of structural brain damage, severe chemical imbalances, and extremely low activity in certain regions of the brain would all appear to be relevant in at least some cases where we are attempting to attribute responsibility to an agent. To take another example, imagine that Carl, normally a law-abiding family man, starts acting impulsively and then gropes a neighbor. He is discovered to have a brain tumor, and when it is removed, he once again becomes considerate and law-abiding. Eight months later, he is arrested for groping a sales assistant. Evidence that Carl's brain tumor has now returned would seem, prima facie, relevant to his responsibility for the assault.

We think the pertinent question is not *whether* brain science can inform responsibility assessments, but in which sorts of cases, and to what extent. If facts about the brain bear on facts about the mind, then neuropsychological findings must in at least some cases inform assessment of defendants' mental states, so long as we can link facts about brains to facts about the capacities necessary for responsible agency. All we need, it seems, is to determine which mental capacities are necessary to responsible agency, and which facts about brains are relevant to those capacities.

This line of thought is disquieting to many prominent philosophers interested in agency and responsibility.[6] Stephen Morse, for example, worries about the spread of "brain overclaim syndrome" (Morse 2006a) into the legal realm. Morse argues that some commentators allow their enthusiasm for the fast-moving world of neuroscience, with its seductive fMRI images and buzzy breakthroughs, to lead them astray. They apply neuropsychological findings incautiously and overestimate their true relevance to questions of criminal responsibility. Though he is not hostile to neuroscience in general, Morse is decidedly skeptical about the evidential significance of neuroscientific results in criminal trials. In his survey of *Roper v. Simmons*, a case in which the U.S. Supreme Court eventually ruled that the death penalty for crimes committed while the defendant was under the age of 18 was unconstitutional, Morse discusses the neuroscientific findings adduced in amicus briefs to the Court:

> Assuming the validity of the neuroscientific evidence, what does it add? The rigorous behavioral studies already confirm the behavioral differences [between adolescents and adults].... At most, the neuroscientific evidence provides a partial causal explanation of why the observed behavioral differences exist and thus some further evidence of the validity of the behavioral differences. It is only of limited and indirect relevance to responsibility assessment, which is based on behavioral criteria. (Morse 2006a, 408–9)

Morse's worries about neuroscience's incursion into the criminal law mostly concern the prospect that overzealous legal scholars will wrongly inflate the significance of neuroscientific findings, or that neuroscience—which is after all a rather young scientific field—may get some things wrong altogether. That is, Morse worries that some are claiming the relevance of brain science to legal responsibility without building an appropriate bridge between the two types of facts. At times, he even seems to worry that such a bridge cannot be built because "the way the brain enables the mind continues to be a mystery" (Morse 2013b, 512).[7] We would caution against thinking of brain and mind as two separate but causally related entities, though, and suggest that a good portion of the mystery derives from this thinking. But luckily, we will argue, we need not enter into such metaphysical disputes in order to develop an effective and sound approach to responsibility grounded in science. Although Morse may be right to worry about certain aspects of the relationship between brain and mind—the mystery of how brain matter gives rise to conscious phenomenal experiences, for example—progress in neuroscience and cognitive science continues to inform our understanding of how the brain enables our cognitive capacities and functions. How a physical system can fulfill certain functions is not a great source of mystery, and we will argue (primarily in chapter 5) that our remarkable abilities to be responsible derive from these cognitive functions, rather than our equally remarkable ability to be conscious. Given this, if we are clear on the capacities that matter to responsible agency, we should be increasingly able to use neuroscience to understand how brain function supports or undermines these capacities as science progresses.

Other scholars have staked out positions at the other end of the continuum from Morse and worry that neuroscience will get it right—all too right, one might say, since they see the relevance of neuroscience to questions of responsibility and the criminal law as not only direct but radically disruptive. Scholars harboring these concerns tend to see our everyday folk-psychological responsibility assessments as deeply committed to a libertarian conception of free will, which is threatened by the findings emerging from brain science.[8] If their view is accurate, then a complete neuroscientific understanding of human agency might well supplant our cherished commonsense notions of free will, agency, and responsibility, which ground our legal practices of blame and punishment. This may then lead to verdicts about responsibility that differ strongly from our commonsense notions of who should be held responsible. A prominent line of conjecture holds that neuroscientific data may require us to shift away from retributive (blame-based) punishment and toward a rationale for punishment that is exclusively concerned with deterrence and/or rehabilitation.[9]

Defenders of this sort of view ignore the proliferation of work articulating compatibilist theories of responsibility, however. A large and diverse group of scholars see the folk intuitions about agency and responsibility as compatible with physical determinism and believe that what libertarian commitments exist can be revised without much disruption to the structure of responsibility assessments.[10] That is, the folk-psychological concepts that underpin assessments of responsibility—where mental capacities and states are attributed to an agent to determine the level of praise or blame that constitutes an appropriate response to a particular action—do not seem to be undermined by the brute fact that these capacities and states are realized in the brain (which is a physical object subject to certain deterministic rules or laws).

For example, the most popular philosophical account of the capacities necessary for responsibility, the "reasons-responsiveness" account made famous by Fischer and Ravizza (1998), understands such capacities as compatible with determinism. On one version of a reasons account, a person is moderately responsive to a reason for action if he or she would have recognized that particular reason as relevant to his or her action in a "close possible world," which means a world we can imagine that is very like our own except for particular details (this is called a counterfactual reasons account) (Vargas 2013). That is, under other similar circumstances the person would have recognized the relevance of the reason to his or her actions, possibly causing the person to inhibit the action altogether. For example, in the actual world, Mary doesn't remember to pick her child up from soccer because she is trying to finish work due by the end of the day. But if, in another possible world, her deadline were to be extended, she would have remembered. This shows she is appropriately sensitive to reasons to pick up her child, even if she forgets under certain conditions. On this sort of view, even claims that a person "ought to have been more careful" or acted differently would seem compatible with the truth of determinism. We will discuss philosophical theories of responsibility and free agency in more detail in chapter 3, but it will suffice to say here that the compatibilist theories advocated by many philosophers leave wide open the possibility that responsibility assessments may be informed by brain science without undermining attributions of praise and blame, or application of punishment, and in so doing depriving us of justified retributive punishment.

In this book we argue for what we see as a moderate position: Neuroscience is both relevant to responsibility and consistent with our ordinary "folk" conceptions of it. Evidence from cognitive science and neuroscience can illuminate and inform the nature of responsibility and agency in specific, testable ways. We are not alone in this view. Neuroscientist and philosopher Adina Roskies, for instance, has conjectured that "neuroscience might enable us to develop a more sophisticated view of responsibility that takes into account both the cognitive demands and the control demands made by intuitive and legal notions of responsibility, and reconciles them with a scientifically informed view of the brain as a physical system that governs our actions" (Roskies 2006, 423).

The majority of our argument in this book constitutes our effort to take up this challenge. Of course, there are existing accounts, especially from philosophers of law, examining the psychological capacities that ground criminal law verdicts (Duff 2004, Moore 1997, Morse 2006b). Our theory's novelty lies in its effort to inform what exactly these capacities are and how they can make us responsible with findings from neuroscience, many of which are recent. The criminal law, as a high-stakes repository of folk-psychological judgments about responsibility, carries an implicit commitment to a view of human agency as basically free and reason-based.

The image of human agency given by contemporary brain science seems irreconcilably different: that of beings whose actions are governed by the mechanistic churn of an immensely complex physical system. Our goal is to show that these two conceptions of human agency are not in fact incompatible. We will offer, in Roskies's terms, a sophisticated view of responsibility that can both serve our folk and legal purposes and be reconciled with and reinforced by a mature scientific understanding of the brain. Hence, our title: *Responsible Brains*.

The Road Ahead

In the chapters to come, we will argue that folk conceptions of responsibility, which underpin and are reflected in the structure of criminal offenses and verdicts, implicitly refer to a particular set of cognitive and volitional capacities. These capacities are implemented by brain structures primarily—though not exclusively—belonging to the cognitive control network and are known to the neuropsychological field as *executive functions*.[11] These functions are what allow us to be agents with reasons, plans, and values and to coordinate our behavior accordingly as we move through a complex world. Executive functions—such as attentional control, planning, inhibition, and task switching—are therefore uniquely well suited to ground a reasons-responsiveness account of the capacities necessary for moral responsibility, including both sensitivity to morally or legally relevant reasons and the volitional control to act in accordance with those reasons.

When we analyze paradigmatic cases of legal responsibility, as well as cases "from the margins" of excuse, as Shoemaker (2015) labels them, we find that legal responsibility is contingent upon a person's having the capacity for some baseline level of executive function, either at the time a crime is committed or for some significant period of time before the crime. We claim the law assumes that citizens possess some baseline executive capacity, and that cases of legal responsibility are cases in which an offender either exercised executive functions with regard to the criminal act or could have done so, in the counterfactual sense described above (unless some justification applies). We will argue, further, that cases of legal excuse tend to be cases where an offender had severely compromised executive functions—because, for instance, he or she suffered from schizophrenia or was too young at the time of the offense to have had a minimally mature set of executive capacities.

Let us map out the road ahead. The early chapters of this book present our executive theory of responsibility, along with its philosophical and scientific foundations. Chapter 2 will give a detailed review of the current neuroscientific research on executive functions—what they are, how they are realized in the brain, and how they are measured and assessed. We will discuss the relationship between the various executive functions, their organization, the extent to which they are separable, and their relevance to reasons-responsiveness. Chapter 3 will motivate our position and place it within the broader philosophical landscape of theories about free agency and moral responsibility. Chapter 4 describes how our theory meshes with legal theory and the criminal law itself.

In chapters 5 through 7, we attempt to further support our theory by arguing its merits compared to those of a prominent competitor: the intriguing consciousness thesis advocated by Neil Levy (2014), according to which consciousness is crucial for responsible agency. We have chosen to discuss Levy's theory as much for its strengths as for what we see as its flaws. His view is nuanced, closely argued, and informed by attention to findings in the cognitive sciences. And indeed, we think Levy is barking up the right tree, because consciousness and executive function are strongly related phenomena. But they are different phenomena. We will argue that it is actually executive functions, which are anatomically and functionally separate from the process of consciousness itself, that do the important work of enabling reasons-responsiveness and, therefore, responsibility.

Having articulated the core of our theory and argued for its merits in comparison to a prominent rival view, we turn in the later chapters to extending and applying that theory. Chapter 6 continues what we see as a fruitful comparison between our theory and Levy's while fleshing out our claim that our theory is folk-consistent, by explicating how it would assign specific truth values to everyday folk claims about responsibility. In chapter 7, we describe two cases in detail—one man who neglected to watch his children over the weekend, and another with a rare sleep disorder who attacked his in-laws in the middle of the night—explaining how our theory's treatment of them is preferable to Levy's. Chapter 8 takes up the puzzle of juvenile responsibility—how, on an executive account of responsibility, the capacities for responsible agency take root and develop in the maturing brain. The gradual maturation of executive functions calls for us to think of juvenile responsibility in scalar terms, with consequent implications for reforming the machinery of juvenile justice. Chapter 9 explores the ways that mental disorders or diseases may undermine responsible agency, applying the executive theory of responsibility to the doctrine of legal insanity and the question of whether psychopaths are culpable for the harms they cause. Chapter 10 discusses the implications of our view for thinking about criminal punishment. If we take seriously the notion that executive function is the key to responsibility, we may need to revise our sense of why punishment is justified in the first place, and how it should be applied to best serve the proper functions of criminal justice. Chapter 11 will briefly recapitulate our main claims, revisit the cases described in the first part of this chapter, and outline some future avenues of inquiry and practical application suggested by our theory.

2 The Science of Executive Processes

Our goal is to build a bridge between our ways of thinking about responsibility and our current understanding of the brain. We will detail a hypothesis according to which certain brain networks perform cognitive functions that play a vital role in allowing people to be responsible for their actions. The bridge we are building in this book is not the material equivalent of a four-lane superhighway, made of concrete and steel. That bridge will come later. Our bridge is made of temporary materials: construction scaffolding (in the form of tentative hypotheses that are likely to be replaced later) and techniques for fastening parts together (logical and causal inferences). Our raw materials, the scientific hypotheses, are strong when used correctly, that is, when used with full awareness of their limits. We will work in both directions: from an account of our folk-psychological sense of responsibility to brain functions and their networks, and in the other direction, from what is being learned and hypothesized about these brain networks back to our everyday and legal notions of responsibility.

Rather than assume it, and take any chance of being misunderstood, in describing scientific results in this chapter we understand that all science is hypothetical, especially recent science and its theory. Some hypotheses just hang around longer than others before they are thrown out or improved. Some of the science we are going to describe is recent science. This means that we are describing a mix of established theory and newer hypotheses on matters not yet as settled. We will be clear about which is which, typically just by using the terms "theory" or "hypothesis" throughout. But sometimes we will take it for granted that the words "According to theory x" should proceed large portions of what follows.

Thus, in this chapter, our claims will often have the following form: *If* this body of scientific theory and experimental evidence is correct, here is one implication it has for our understanding of the human capacity for responsibility. In this chapter, our goal is to convince you that there is strong scientific evidence for the existence of executive processes, and that the notion of executive processes will itself become an important part of the neuroscientific theory of the brain. Perhaps a new theoretical perspective on what the prefrontal lobes are doing will emerge, and people will move away from executive theories at some point in the future. We are betting not, or at least not anytime soon.

Our best current overall picture of brain function from neuroscience is that there is a fundamental distinction in the way the brain treats routine as opposed to novel situations. Routine situations, such as a daily, unchallenging drive to work, do not require decision-making or action planning. We just get in the car and go "on autopilot" while engaging in routine actions. But imagine that one day, on the way home, you encounter a roadblock due to construction. Now novel actions are required: You must find a new way home, possibly a route you have never taken before. Decisions need to be made about which possible route is best. Some contemplated routes will need to be ruled out; for instance, you remember that a certain road will have heavy traffic today. A plan needs to be made—which turns need to be made where, which lanes you should be in, when extra vigilance or care will be required, and so on. You can't relax until you are almost home since you are looking for street signs, struggling to negotiate lane changes in heavy traffic, recalculating your route based on a missed turn, and whatever else may come up. You arrive home exhausted by all the mental effort.

These sorts of novel, complex tasks recruit an organized set of processes that form a network located primarily in the brain's prefrontal lobes known as executive processes or cognitive control processes. Creatures without such processes have little behavioral flexibility. Spiders, for example, despite their ability to exhibit complex behaviors such as web-building and prey capture, do not make decisions. They are always on autopilot. By contrast, human beings' robust suite of executive functions allows us not only to behave flexibly, but to exert self-control—to guide our actions in complex ways, taking account of input from our senses, or from memory, but also from our emotions and values. Executive functions allow us to plan actions and formulate intentions, to set and pursue goals, to organize complex actions with multiple parts and phases, while not losing our place, and to establish or revise patterns of habitual behavior.

Executive functions are distinct from other mental capacities, such as perceptual capacities, both with regard to what they do and where they reside in the brain. Neuroscientists draw a rough but defensible line in the brain between, on the one hand, perceptual processes and, on the other hand, executive processes that can correct perceptions and formulate intentions based on perceptual input. Sensory processing begins in the back of the brain and then steadily moves toward the front. As processing continues, several functions are accomplished: Certain details of the incoming sensory flow are highlighted and emphasized; the boundaries of visual objects are made sharper; in the auditory realm, the boundaries between words are sharpened; and missing information is filled in, as in the case of the visual blind spot. When the sensory systems are finished, highly processed representations are made available to the brain's evolutionarily newest parts, the prefrontal lobes. However, all this work is of no use if the representations are not employed in an intelligent and effective way. This is where executive processes come in. They achieve what we speak of in our everyday mental talk as planning, thinking, decision-making, and even reasoning, the ultimate goal of which is to conduct ourselves in a way that meets our objectives effectively, within a complex social milieu.

The same story we just told about how perception relates to the executive processes can be told about memory and emotion. Memory and emotional processing are functionally and anatomically separate from executive processing although they interact closely. The memory system consists both of areas devoted to storing and reactivating memories, but also of prefrontal areas that are able to check putative memories for accuracy and reject false memories. Executive processes are also responsible for managing the effect of emotions on both thought and behavior; although emotions in general are generated outside of the prefrontal cortex, their effects on our actions are carefully filtered by the set of executive processes. The most obvious example of this is a case in which we need to prevent a strong emotion—for example, anger—from causing behavior that would be inappropriate in our current social setting. We need to inhibit those angry actions. Current large-scale brain theory in neuroscience continues to make and employ clear distinctions between sensory, mnemonic, and emotional networks and their constituent areas, on the one hand, and executive network areas devoted to cognitive control on the other. Ideally, the executive processes take all three—current perceptions, along with relevant memories and emotions—into account in formulating intentions to act and then monitoring and guiding the subsequent behavior.

Our control over our abilities to perceive, remember, and emote has certain concrete limits, which is itself another sign of the fundamental separateness of these abilities from the processes that do the controlling. A color-blind person, for example, cannot simply decide to start seeing colors correctly. An elderly person with a bad memory cannot simply decide to start remembering. A sociopath who fails to feel empathy for the people he harms cannot simply decide to start feeling empathy. However, our control over the actions we take on the basis of our perceptions, memories, and emotions is vastly greater. Once the color-blind person knows that he sees both red and green as green, he can take steps to ensure that his actions do not endanger himself or others. For instance, even though red and green traffic lights may look the same to him, he can memorize the position of the lights—red on top, green on the bottom—to make sure that he does not cause an accident. He can use executive processing to correct for a perceptual deficit.

Our everyday assessments of responsibility reflect this basic division between perception, memory, and emotion, on the one hand, and the realm of thought and decision-making on the other. If the color-blind man drives through a red light and causes an accident, we will hold him responsible, because he could have corrected for his color blindness but failed to do so. We hold people responsible for actions that arise from failures to correct certain problems when we believe they could have done so. Echoing this folk sentiment, one of the primary claims of this book will be that only brain damage—or any sort of suboptimal brain functioning—that causes deficits that cannot be corrected for by the executive processes are grounds for moral or legal excuse.

To take one example of potentially exculpatory brain dysfunction, in chapter 9 we will discuss in detail the case of individuals with schizophrenia. We will argue that certain schizophrenics may be excused from their violent acts because they can no longer identify and correct for hallucinations and violent impulses to cause harm. Schizophrenia attacks brain function at multiple levels, and when it affects the executive processes seriously enough, it can render the person incapable of taking note of hallucinations and delusions as carrying false information about the world. The serious executive dysfunction often brought about by schizophrenia (see, e.g., Orellana and Slachevsky 2013, Minzenberg et al. 2009) compromises a person's abilities to reason, preventing the person from being able to reject paranoid delusions as well as desires to do violent harm to others. Thus, schizophrenics can have deficits both in perception—they are seeing or hearing things that are not there—and in executive function, which in healthy brains allows one to identify faulty perceptions and inhibit acting on them. The combination of the two sorts of deficits may be at the root of the schizophrenic's harmful behavior, but it is the higher-level executive deficits that are exculpatory, according to our view.

The Theory of Executive Processes

The existing theory of executive processes draws on a diverse group of brain sciences, including neuroanatomy and physiology (the study of the brain's anatomy, especially its connectivity, and brain function), cognitive neuroscience (e.g., brain imaging experiments, in which subjects are given tasks requiring executive intervention while their brains are imaged), cognitive psychology (e.g., perceptual experiments involving behavioral responses by subjects), and behavioral neurology (neurological patients who have prefrontal lesions are studied to determine which mental functions have been compromised by the damage, e.g., Phineas Gage [Harlow 1868] or Antonio Damasio's [1994] patient EVR).

Initially it was thought that while the different executive functions tend to involve the prefrontal cortex, they were accomplished by different cortical areas. Imaging studies began to find, however, that different executive tasks recruited similar sets of cortical areas (Duncan and Owen 2000a). At the same time, there was difficulty establishing clear cases of double dissociations between executive functions (see, e.g., Duncan and Owen 2000b). A double dissociation between processes x and y occurs, for example, when a patient has executive capacity x but lacks y, while another patient has y but lacks x. While not conclusive, such double dissociations provide evidence that the two functions are performed by two different brain processes.

More recent research, however, appears to show that all executive functions (or at least a core set of them) are accomplished by a single, unified brain network, the frontoparietal cognitive control network (working together with adjunctive areas, some of which are unique to the particular executive process involved). The idea of a cognitive control network, which is a neuroscientific concept about a set of cortical areas and supporting areas that have been shown to work as a system, ties the theory of executive functions to neurobiology, in that it specifies what biological processes achieve those functions. "Cognitive control network" is the name of a physical set of interconnected brain area and organs. This bodes well for approaches such as ours since it indicates that the idea of executive functions has a chance to be an enduring one as the cognitive sciences move forward.

Dysexecutive Syndrome

One way to get clear on executive processes is to look at what happens to people who lose them. During the twentieth century, neurologists and other scientists began to accrue cases in which patients with frontal damage displayed an intriguing pattern of higher-level deficits. If you encountered such a person, unless your conversation got deep or lengthy, you might well never know you are dealing with someone with significant brain damage. These patients can register normal IQ scores and perform routine actions easily. But in their lives, "they are impulsive and irresponsible and consequently can have trouble holding a job, remaining married, and so on" (Miller 2000, 61). Their brain damage makes them creatures of the moment. As Miller puts it, "[T]he behavior of humans and monkeys with prefrontal damage can be described as stimulus bound. Their behavior is captured by salient sensory cues that reflexively elicit strongly associated actions. They are unable to override these impulses to engage in behaviors that depend on knowledge of a goal" (Miller 2000, 61). People in this condition merely appear normal. They cannot function at a job, or in society in general.

During the ensuing decades, however, patients were found who showed similar symptoms, but as a result of damage to nonfrontal parts of the brain, such as the parietal lobes. In the 1980s Alan Baddeley proposed that the pattern of behaviors constituted a genuine syndrome, which he called the dysexecutive syndrome (Baddeley and Wilson 1988), which was functionally rather than anatomically defined (Baddeley 1998, 2002). Baddeley and Wilson's original patient had poor recall of autobiographical memories and tended to confabulate, by producing false autobiographical memories. Patients without damage to executive processes, on the other hand, do not confabulate and will admit that they do not remember (Parkin 1984, Moscovitch and Melo 1997). As the syndrome is described, dysexecutive patients, "despite the superficial appearance of normality ... have difficulty in sustaining attention, in keeping on task, and seem to act on whims and impulses without regard to future consequences" (Miller, Freedman, and Wallis 2002, 1124).

Just as there is an unresolved question as to how exactly to divide the executive functions, there is a matching issue about whether there really is a general dysexecutive syndrome or whether the executive processes can fail separately. Stuss and Alexander (2007), for example, argue that three different executive functions map onto three different prefrontal areas that can be damaged independently of one another. One way to account for the dispute is as follows: The brain networks involved in the cognitive control network (see below) seem to have two or three crucial nodes in addition to several adjunctive nodes, where activity increases only with certain executive tasks, for example, these areas are active only during planning, not during any other executive activity. This may shed light on the nature of the loss of executive function: General loss might follow from lesion of one of the core frontoparietal areas. Assuming that the rest of their brains are healthy, people with this sort of damage would tend to fit the general dysexecutive profile. But also, there could be specific losses resulting from damage to an adjunctive area necessary for a certain executive function. This new neuroscientific picture of executive function may well explain the difficulty earlier researchers had in establishing cases in which brain damage affected only a single executive process (Duncan and Owen 2000b). The approach also explains why subjects' scores on different tests of executive function are correlated (Gilbert and Burgess 2008).

A List of Executive Functions

There is a plausible new view according to which the different executive functions are achieved by the cognitive control network, a small set of interconnected core areas, in conjunction with a larger set of adjunctive areas (Niendam et al. 2012, Lindquist and Barrett 2012). Any particular executive function is thus supported by the set of core areas (see below) working together with a subset of the adjunctive areas, but the latter subset is different for each executive function, albeit with some overlap. The list of executive functions that are accomplished by the cognitive control network shows variation among different researchers. What we have done below is to include functions from several different authoritative sources, and group them into what seem to us to be natural categories. We expect the list to continue to evolve as research progresses.

Attention

Alternatively, some authors speak of *vigilance* (e.g., Niendam et al. 2012) while others speak of *monitoring*. As is typical with an executive process, we find it effortful to maintain attention. A core deficit of the dysexecutive syndrome is a short attention span, so that patients are unable to follow a story line or keep to a conversational topic (Wilson et al. 1998). There are both unimodal and multimodal attentional networks. Top-down attention is considered to be under voluntary control, as opposed to bottom-up attention, in which our attention is involuntarily drawn to movement in the periphery of the visual field, for example (Posner 2011). Attention can be particularly important to generating socially appropriate behavior: If one doesn't notice social cues, for example, one cannot respond to them. In a more extreme case, if a driver doesn't notice the traffic light, he is not going to stop.

Monitoring of Perceptions, Memories, and Emotions

Perceptions cannot simply be taken at face value. They often need to be checked against our memories or against other perceptions. For instance, I believe I see the person I am looking for in a crowd, but closer attention reveals that it is merely someone who looks like her. Similarly, in the case of memory, Baddeley (1998) lists "interfacing with" long-term memory as an executive function. Certain memory patients are also a good example of this corrective ability. As we noted, people with damage to the memory system, but intact executive processes, admit their memory problems and take steps to correct for them. They might report memories with caveats, such as, "It seems to me that I remember…" or "That's how I remember it, but I have a bad memory." These memory patients are like the color-blind driver in that they are able to correct for their deficits in a way that allows them to continue to exist in society. But people with damage to both memory systems and to the relevant executive systems will deny that they have a problem and produce confabulations (Hirstein 2005). Some of these confabulations can have disastrous consequences for the person who issues them, but also for others. However, once we know about their executive deficits, we tend not to hold confabulators responsible for their false claims.

Monitoring of Behavior

"Monitoring is the process of checking the task over time for 'quality control' and adjustment of behavior" (Stuss and Alexander 2007, 909). Attention and monitoring during behavior seem particularly important for culpable action. Executive processes activate when extra attention is needed during an action because it is no longer routine, when there is danger, or when the stakes are high. Intentional actions in general are monitored during their execution, and this monitoring is a type of attention, that is, paying attention to what you are doing. A complete failure to monitor can be a sign that the action is not intentional, that the person is sleepwalking, hypnotized, having a seizure, and so on. During monitoring, executive processes causally interact with ongoing conscious states and can inhibit subsequent intended actions and activate planning processes. Attention and monitoring can allow for a type of causally continuous executive control over actions. These monitored actions are the opposite of routine actions, and also different from actions in "flow" states, which we will argue also lack executive involvement.

Working Memory

Speaking of working memory as an executive *function* is counterintuitive, based on the idea that the brain's memory areas are mnemonic stores or buffers, rather than functions. This would make it more of an executive resource than a function. Typically though, when neuroscientists list working memory as an executive function, they are treating the mnemonic store together with the areas that manipulate it and update it, which are more plausibly thought of as executive functions (Milner 1982, Goldman-Rakic 1996, Levy and Goldman-Rakic 2000, Curtis and D'Esposito 2003).

Planning and Goal Setting

Planning is perhaps the clearest example of an executive process, and it was one of the first executive processes to be traced to the prefrontal lobes (Owen et al. 1990, Pennington and Ozonoff 1996, Smith and Jonides 1999, Shallice 1982, 1988). Actions that are important or complex require careful forethought, including the formulation of alternatives that may be needed, as well as attention to the exact order in which tasks must be performed. Miller and Wallis (2009) mention the patient with prefrontal damage who first stirred her coffee and then added milk. Long-term plans can become complex enough that they have to be broken down into primary goals and subgoals, something only executive processes can accomplish. To take an example, the process of choosing and becoming qualified to get a certain sort of job (e.g., a lawyer or a plumber) or to become a certain sort of person (e.g., a healthy person or an honest person) often takes years of meeting subgoals before the ultimate goal is reached.

People with dysexecutive syndrome do not make plans for the future. They may spend all their money on silly things, for instance. They also cannot solve more complex problems that require a plan. Another intriguing symptom shown by some more severe dysexecutive patients that shows how environmentally bound they are is called utilization behavior: They will pick up and use whatever is put in front of them. Put a comb in front of them, and they pick it up and comb their hair. Don't put your coffee cup down near them; they will pick it up and drink from it. The patient retains the habituated behavioral routines of combing hair and drinking coffee but can no longer exercise top-down control over those routines and fit them with intentions and larger contexts.

In addition to planning, some writers assign an executive process called initiation (Smith and Jonides 1999, Lezak 1995) to start the behaviors that will realize the plan. Go/no-go experiments involve a decision about whether to act or not, based on the current stimulus. Stuss and Alexander similarly speak of "energizing," which they say is "to plan, and to activate and sustain the intention" (Stuss and Alexander 2007, 911).

Inhibition

Several different types of processes at several levels of brain function are referred to as inhibitory, or as effecting inhibitions (Leimkuhler and Mesulam 1985). Some neurotransmitters are excitatory while others are said to be inhibitory, for example. The inhibitions effected by executive processes lie at the other end of the hierarchy of brain theory. Miyake et al. (2000, 57) say that inhibition involves the ability to "deliberately inhibit dominant, automatic, or prepotent responses when necessary." The ability takes time to develop during childhood and adolescence (Luna, Padmanabhan, and O'Hearn 2010). Damage to the systems responsible for inhibition can result in perseveration, in which individuals continue with a course of action they realize is wrong but cannot stop themselves from engaging in.

Disinhibition is a neurological condition that occurs when people lose the executive capacity to stop unwise, illegal, unethical, inappropriate, or just ineffective actions from being executed. It can be seen in the extreme form in the elderly man at the retirement home who throws off all his clothes and runs down the hall, but it exists in much more subtle forms. We are all guilty of at least briefly considering all sorts of unethical behaviors. The executive process of inhibition needs to stop these thoughts from continuing to form into intentions to act and, finally, into actions themselves. There is no single inhibition area in the brain. Rather, what one sees is that inhibition in different sorts of tasks involves different, although sometimes nearby, brain areas. People who are diagnosed with disinhibition are generally not considered responsible for their actions. Also, as Damasio (1994) showed, there are complex, high-level inhibitions that require a delicate combination of gut-level, somatic responses, educated by higher-level executive functions that are capable of taking additional factors into account.

Task Switching

Many complex actions require that we move back and forth from one action to another, while keeping our place in each of the two actions (Braver 2001, Ravizza and Carter 2008, Pashler 2000). For example, when I interrupt my writing to check on an odd noise outside, I am able to quickly take up where I left off. Some researchers speak instead of flexibility (e.g., Niendam et al. 2012); a lack of cognitive flexibility had been noticed by early researchers of the effects of frontal lobe damage (e.g., Milner 1982). Others speak of dual-task performance (Baddeley 1998) or mental set shifting (Hofmann, Schmeichel, and Baddeley 2012).

Broad Descriptions of the Set of Executive Processes

Another way to understand executive function is to survey some descriptions of what the set of executive processes, taken as a whole, accomplish. According to Banich, "[T]he executive functions can be thought of as the set of abilities required to effortfully guide behavior toward a goal, especially in nonroutine situations" (Banich 2009, 89). The executive processes manage both thought and behavior in ways that allow us to accomplish our goals.

Neurons in the prefrontal cortex are "activated by stimuli from all sensory modalities, before and during a variety of actions, during memory for past events, in anticipation of expected events and behavioural consequences, and are modulated by internal factors such as motivational and attentional state" (Miller, Freedman, and Wallis 2002, 1124).

Miller and his colleagues speak of the prefrontal cortex as containing a special type of representation—a plan: "[T]he PFC rapidly constructs a pattern of activity that represents goals and the means to achieve them" (Miller, Freedman, and Wallis 2002, 1132). The prefrontal cortex then enhances "the activity of neurons representing task-relevant information, [while] those representing task-irrelevant information are simultaneously suppressed and neural activity is steered down the pathways needed to solve the task at hand" (1133). In the service of this, prefrontal neurons show "conjunctive tuning for learned associations between cues, voluntary actions, and rewards" (1133).

Some authors include decision-making in their list of executive functions. But the process of decision-making seems too complex to be considered an executive function by itself. Decision-making can be broken down into several executive functions, such as planning, engaging in conflict resolution, inhibiting, and so on. However, the term "decision-making" might work better as an attempt to describe what the executive processes do in general. This fits with the idea that executive processes activate when decisions are needed because routine behavior failed or will fail. Not all executive activity seems correctly describable as decision-making though. If it is right to call the little inhibition necessary to state the colors and not read the words in the Stroop test an executive process, we don't believe that we want to also commit ourselves to there being a little decision each time, to inhibit reading, attend to color, and so on.

Notice that both descriptions above focused on the notion of planning to achieve goals as central to characterizing executive activity. The executive processes plan actions and then initiate and monitor those actions according to the plan. Planning requires attention to perceptual and mnemonic information; it involves inhibition, when a routine response must be stopped or when a candidate plan is inhibited because its simulation resulted in (represented) negative consequences. Testing candidate plans requires working memory, as well as other representational systems. A complex plan will include prescriptions for attention and monitoring certain properties at certain times; it will also include certain possible "inhibition points," where one should take care not to do certain things, and it will include certain choice points, where flexibility is required.

Planning also requires the involvement of the brain's emotion and reward systems. I need to assess the reward value to me of different possible outcomes of a plan. I also need to access my memories of what sorts of plans and actions were rewarded or punished in the past. I may also need to use my brain's representational resources, such as the default mode network's (see below) ability to represent certain possible or future situations. If a plan passes all the tests and the time is right to put it into action, intentions must be formed for the first actions involved in the plan, and those actions must be initiated. The actions must then be monitored and compared to the plan and its expected reward until the plan is complete. This planning-based approach to the executive processes honors the truism that the human prefrontal lobes grew in size during our evolutionary history in order to allow us to perform much longer perception–action cycles (Fuster 2002). Rather than being trapped in the present like simpler organisms, we can transcend the present by keeping a careful record of the past, but also by using our abilities to envisage future events or possible events.

The Cognitive Control Network

Tara Niendam and her collaborators (2012) conducted a meta-analysis of 193 functional neuroimaging studies on more than 2,800 healthy adult subjects as they performed executive tasks. They found that the full range of executive tasks employed in these studies activated a core set of areas, which includes dorsolateral prefrontal, dorsal anterior cingulate, and parietal cortices (see figure 2.1, plate 1, as well as the orange areas of figure 2.2, plate 2; see also Cole and Schneider 2007; Collette et al. 2005). In addition to these core areas, different executive tasks activated connected adjunctive areas, including anterior prefrontal cortex, anterior midline and midcingulate areas, together with subcortical areas such as the basal ganglia and cerebellum. We will refer to this network, consisting of both the core and adjunctive areas, by its functional name, as the cognitive control network or, interchangeably, by its anatomical name, as the frontoparietal network.

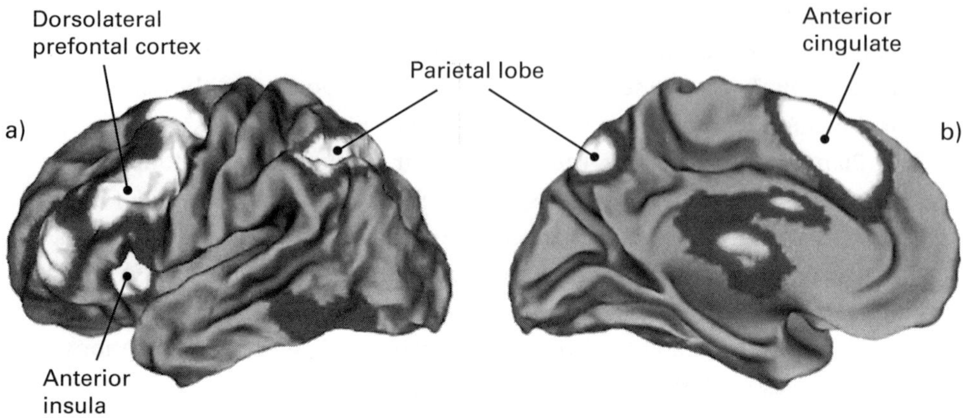

Figure 2.1
Satterthwaite et al. diagram summarizing findings on adolescents performing a working memory task, the n-back task, which "robustly recruited the executive network" (Satterthwaite et al. 2013, 16253). Subjects are shown stimuli one at a time and must decide if the current one is the same as the one, e.g., 2-back. (a) Left hemisphere, lateral view. (b) Left hemisphere, medial view. Adapted from figure 4, p. 16253, of Satterthwaite et al. (2013). See color plate 1.

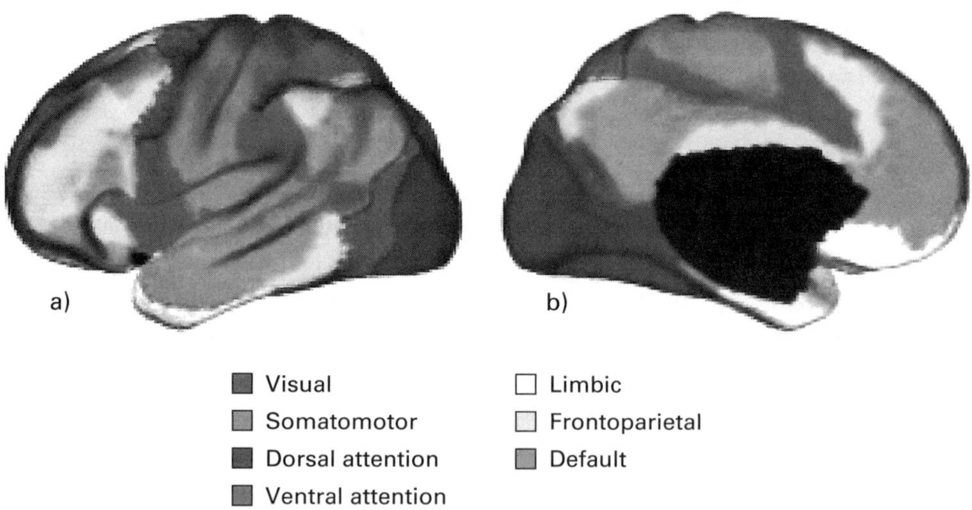

Figure 2.2
Summary map of Yeo et al., showing the extent of their seven cortical networks. The cognitive control network is in orange and is referred to here as the frontoparietal network. (a) Left hemisphere, lateral view. (b) Left hemisphere, medial view. Adapted from figure 11, p. 1137, of Yeo et al. (2011). See color plate 2.

Constituent Areas of the Cognitive Control Network

Bilateral Dorsolateral Prefrontal Cortex (Brodmann Areas [BAs] 9, 46) The dorsolateral prefrontal cortex[1] has connections to sensory and motor regions that allow it to maintain rules for action, response selection, and inhibition (Watanabe et al. 2002). This cortical area also supports the ability to shift attention depending on how well the goals of the task are being met. Petrides (2000) argues that the primary function of dorsolateral cortex is to monitor and manipulate items in working memory. Frith (2000) points out studies showing that the dorsolateral cortex is highly active when subjects must choose from among alternatives, arguing that this indicates that its function is to create possible responses in nonroutine situations by biasing cells in posterior cortex that represent the responses.

Dorsal Anterior Cingulate Cortex (BA 32, Medial Area 6) The anterior cingulate cortex, along with nearby medial frontal regions, detects conditions, such as response conflicts (Carter et al. 1998, Milham et al. 2001) that require executive control, and performs other types of monitoring (Luks 2002). It then activates the dorsolateral prefrontal cortex (Egner and Hirsch 2005), which then participates in the cognition needed to resolve the conflict. This cognitive or executive portion of the cingulate should be distinguished from nearby cingulate areas that have connections to the brain's emotional systems (Bush, Luu, and Posner 2000).

Parietal Cortex (BA 7) The parietal cortex seems to be playing a role in providing the dorsolateral cortex with "information on stimulus salience and learned stimulus–response pairings" (Niendam et al. 2012, 256, Bunge et al. 2002). A meta-analysis of imaging studies on subjects performing a classical working memory task, the n-back paradigm, showed frontal areas active along with parietal and cerebellar areas (Owen et al. 2005). Parietal areas also contribute to certain types of cognitive flexibility. Ravizza and Carter (2008) found parietal activity when subjects switched attention between different features of a stimulus. Superior parietal cortex (BA 7) activity was seen by Niendam et al. (2012) as an area of aggregation among studies, as was activity in precuneus/occipital lobe (area 19), and the anterior insula (see figure 2.1), which researchers have argued may function to alternately connect the cognitive control network with other brain networks (see below).

Adjunctive areas that also showed up regularly in the Niendam et al. meta-analysis include premotor cortex (area 6), a complex of areas having to do with performing motor acts as well as understanding the motor activity of others; frontopolar cortex (area 10), which appears to play a role in multitasking and high-level planning (Petrides 2005, Burgess et al. 2000); and orbitofrontal cortex (area 11), which can supply information about past reward values associated with different stimuli. Executive tasks can also involve activity in temporal cortex (areas 13, 22, and 37) when auditory stimuli are processed and in occipital cortex (area 19) when visual stimuli are processed. Subcortical structures have also been shown to be active during executive tasks, including the thalamus, caudate, and putamen. Portions of the cerebellum, including the posterior declive and the anterior culmen, have also been seen to be active during executive activity (Bellebaum and Daum 2007).

Portions of the overall network show different levels of activity depending on the executive task the subject is undertaking. Vincent et al. (2008) showed that the cognitive control network can flexibly engage with either the default mode network (described just below) during autobiographical planning or the dorsal attention network (see below) during visuospatial planning. Koziol et al. (2014) attempted to distinguish the contributions of the left-hemisphere frontoparietal network from that of the right-hemisphere frontoparietal network: The variant in the left hemisphere guides behavior according to the current "cognitive context," that is, what one is currently thinking about (Koziol et al. 2014, 242), they claim. In contrast, the right-hemisphere variant "is critical for cognitive selection driven by the external environment and for context-independent behavior" (Koziol et al. 2014, 242). The right-sided network also specializes in solving novel problems, according to Koziol and his colleagues.

Additional Brain Networks

Learning the functions associated with different cortical areas is a long and difficult task, undertaken by students in the brain sciences. But an emerging approach simplifies this task by grouping cortical areas into functional networks. Yeo et al. (2011) parcel the human cortex into seven networks: visual, dorsal attention, ventral attention, somatomotor, limbic, frontoparietal (i.e., cognitive control), and default (see figure 2.2). There are three visual networks, including an early network consisting of the occipital lobe, the lateral temporal cortex, and the superior parietal lobule (purple areas in plate 2). This network interacts with both ventral (violet areas) and dorsal attention networks (green areas). The dorsal attention network has traditionally been known as the dorsal stream, or the "where" stream. We understand the attempt to replace the more standard terminology of the *what* and *where* streams with something more formal sounding, but there is a problem with speaking of the dorsal network as an attentional network. The problem is that we are not in general *attending* (in the "top-down" sense) to what this network is doing. In the case of vision, the dorsal stream is fed more by information in the visual periphery, while the ventral stream primarily receives incoming visual flow from the visual focus. On the other hand, it seems that attention can move into the visual periphery. In general, we look at what we are visually attending to, but not always. It is possible to visually monitor someone who is in the periphery of your vision, perhaps because you specifically don't want to look directly at them. More on these issues in chapter 5.

There is also a sensorimotor/somatomotor network (plate 2, blue areas). Constituent areas include primary motor cortex, primary and secondary sensory cortices (anterior regions of the parietal lobes), the supplementary motor cortex, the ventral premotor cortex, the putamen, and portions of the thalamus and the cerebellum (Koziol et al. 2014).

The Default Mode Network

Another large brain network was accidentally discovered when brain imaging researchers noticed that it tended to activate in their subjects between experimental trials when the subjects were presumably thinking random thoughts of their own (Raichle et al. 2001). This network becomes active when we are thinking about ourselves in certain ways, when we think about our futures, when we work to understand the minds of others, and when we recall autobiographical memories (Buckner, Andrews-Hanna, and Schacter 2008). Other candidate functions include self-referential cognition and affective cognition (Gusnard et al. 2001). The network functions as a general purpose representational resource that is employed by the cognitive control network to solve various sorts of problems. Core areas of the default mode network include medial parietal cortex, portions of the lateral frontal lobes, the posterior cingulate cortex, the temporoparietal junction, and portions of the medial temporal lobes (see the red areas of plate 2).

Andrews-Hanna and her colleagues (Andrews-Hanna 2011, Andrews-Hanna et al. 2010) were able to detect two separate subsystems in the default mode network. The medial temporal lobe subsystem consists of portions of retrosplenial cortex, ventral medial prefrontal cortex, and the posterior inferior parietal lobule. Barrett and Satpute (2013) suggest that this subnetwork is more engaged when "more embodied, 'first-person' information is required to make a judgement (e.g. bodily sensations or feelings related to homeostasis)" (368).

The dorsal medial prefrontal subsystem is made up of the dorsal medial prefrontal cortex, the temporoparietal junction, the lateral temporal cortex, and the temporal pole. Barrett and Satpute suggest that this network is involved in "creating situated conceptualization when abstract 'third-person' information is required to make a judgement (such as knowledge about mental state categories, scene construction, etc.)" (Barrett and Satpute 2013, 368). During future-directed thinking, both subsystems engage simultaneously "presumably to facilitate construction of mental models of personally significant events" (Andrews-Hanna et al. 2010, 550).

Far from being the idle "mind wandering" network some still call it, the default mode network is emerging as a vital resource for solving problems, understanding others, and envisaging events, whether remembered or merely imagined.

The Salience Network

The salience network marks stimuli as important.[2] Barrett and Satpute (2013) divide this network into two subnetworks: The dorsal subnetwork consists of the dorsal anterior insula, and the dorsal anterior cingulate, which are "especially important for selecting motor actions (i.e., response selection)." (Barrett and Satpute 2013, 362). The ventral subnetwork consists of the ventral anterior insula, and the pregenual anterior cingulate, which are "important for directing selection in the 'visceromotor' (or autonomic nervous system) responses that support motor actions and that maintain homeostasis.... The strength of intrinsic connectivity within the ventral 'salience' subnetwork is uniquely correlated with the intensity of affective experience, whereas connectivity within the more dorsal subsystem is uniquely correlated to motor control and changing mental sets" (Barrett and Satpute 2013, 366).

The limbic system also plays important roles in making stimuli salient. There is a limbic network (see plate 2, the cream-colored areas), which is partly cortical and which "generates the motivational valence for engaging in certain behaviors and avoiding other activities, which is essentially decision making" (Koziol et al. 2014, 241). There are two hubs in this network, forming two subsystems: the dopaminergic reward system and the amygdala. The dopaminergic reward system includes the basal forebrain. The amygdala is primarily associated with the noradrenergic system.

Interactions between the Cognitive Control Network and Other Networks

The cognitive control network needs to interact with the brain's salience and reward networks to do its work. Attributing emotions to oneself, for instance, involves activation of the cognitive control network, the default mode network, and the salience network (Satpute et al. 2012). The cognitive control network engages with the default mode network during certain types of cognition. They appear to operate alternately, however, since activity in the cognitive control network has been shown to suppress activity in the default mode network. Sridharan, Levitin, and Menon (2008) argue that a network composed of areas in the anterior cingulate and the right frontoinsular cortex is responsible for switching between the two networks. Spreng et al. (2010) similarly suggest that the anterior insula works as the switch to selectively couple the cognitive control network with either the default mode network or the dorsal attention network. The insula has previously been shown to suppress default mode activity (Menon and Uddin 2010).

But how exactly does the cognitive control network selectively couple with the other networks in order to accomplish tasks? One possible example involves the way that the control network can selectively couple and uncouple with two different, "mutually exclusive" networks. The dorsal attention network and the default mode network are largely segregated from one another. However, the cognitive control network has strong connections with both of them. Spreng et al. (2013) identified three different types of nodes within the cognitive control network: (1) nodes aligned with the dorsal attention network, (2) nodes aligned with the default mode network, and (3) nodes aligned with both networks. They hypothesize that the function of these nodes is to coordinate interactions between all three networks. Those nodes allow the cognitive control network to flexibly couple with either network, depending on task demands (Spreng and Schacter 2011, Spreng et al. 2010).

One region that Spreng et al. (2010) hypothesize to be a "connector region" is the left superior frontal gyrus, which has strong functional connections to the rest of the default mode network but is also connected to several portions of the cognitive control network, including having direct connections to the left middle frontal gyrus (area 6) and medial superior prefrontal cortex. The primary locus of connection between the cognitive control network and the dorsal attention network may be the bilateral dorsolateral prefrontal cortex, Spreng et al. (2010) hypothesize. As would be expected, these regions show no connection to the default mode network. Regions of the cognitive control network that are aligned with both the default mode network and the dorsal attention network include the bilateral posterior lateral middle frontal gyrus (area 6) and the medial superior prefrontal cortex.

Some of the deficits in dysexecutive syndrome do not occur among the standard lists of executive processes—theory of mind deficits, for instance. One possibility here is that brain damage has not affected the default mode network itself, but rather, the damage to executive areas prevents them from interacting with the default mode network. An executive deficit in managing a representational system can prevent that system from contributing to cognition.

In this regard, we might consider our special abilities to understand each other's minds to have a top-down and bottom-up dimension in relation to the cognitive control network, in much the same way that attention comes in top-down and bottom-up varieties. In their bottom-up mode, mind-reading capacities might function like a special type of perception. They can provide us with information about people's goals and intentions by operating spontaneously. In their top-down mode, mind-reading processes can be activated and engaged by the cognitive control network, when we explicitly and effortfully ask ourselves: How will she react to this? We can then edit our plans and intentions based on our simulation of the other.

Some of the recent findings reported here give rise to an interesting new objection. Some might argue that folk psychology does not mesh well with the new approach according to which the cognitive control network accomplishes the executive functions, since the new approach is not able to assign specific brain processes to each folk-psychological concept. Rather, mental functions such as planning and decision-making are accomplished by a number of domain-general processes working in various combinations (see, e.g., Lindquist and Barrett 2012, especially box 1 in their appendices). Even in this case, the different executive functions may still have unique neural loci, but they are different "modes" of the cognitive control network. As we see it, the important issue is whether we can justifiably make claims such as "Planning is accomplished by these brain processes," as we will do. We doubt that many people ever thought that there would be a completely unique planning center or process in the brain. The set of brain areas behind each folk-psychological function may well be unique, however, since as we saw, the cognitive control network can enlist unique combinations of other cortical areas as it performs the different executive tasks.

Measuring Executive Function

The classical tests of frontal lobe functions, such as the Wisconsin Card Sorting Test and the Towers of Hanoi task, have been well studied by numerous disciplines of the mind. The brain areas they tend to activate are also well-known. These classical tests have now been augmented by large numbers of new tasks, vastly increasing the variety of executive functions that can be tested. Various sets of these separate tests have also been amalgamated into larger test batteries, designed to provide an assessment of overall executive function, some of which have passed tests of their basic validity.[3] We believe the existing test batteries would be a good starting point toward developing a measurement tool sensitive to our variable of interest: capacity for responsible actions.

Further research on this topic will proceed along several parallel lines of research, including anatomical, physiological, and behavioral approaches, with each approach working to cross-fertilize the others. In the future, in cases where it is important enough to determine competence, we imagine multiple approaches would be used that would ideally be mutually confirmatory. Prior to this, raw correlations between certain phenomena, for instance, prefrontal thickness and performance on the Wisconsin Card Sorting Test, have been or will be attained, to provide a context for interpreting the findings in any particular case. In one type of ideal case, we would have developmental testing results that could show a clear point at which a person's executive profile altered dramatically, perhaps due to some brain disease. Again ideally, the brain disease would have affected the person's behavior in ways known to correlate with the type and severity of that particular brain disease. The disease might have even shown up in the defendant's life in obvious ways, such as being fired from multiple jobs.

We believe that those devising executive tests should pay more attention, however, to some of the more subtle abilities that come into play during reasoning and decision-making. The ability to estimate quantities, both discrete and continuous, has been offered as an executive measure (see, e.g., Gansler et al. 2014). A subtler version of this ability involves the capacity to attach rough plausibility ratings to claims and reasons for claims. Without this, the human mind seems to be prone to a number of foibles, including self-deception, as well as the fallacy of false equivalency, because the person cannot weigh the difference between say, an accidental nudge and a deliberate punch, or a strong argument and a weak one.

With more sensitive measures of executive function, we may also find that certain types of people who were thought to have intact executive functions do not. For example, psychopaths, especially the "successful" ones, that is, those not entangled in the legal system, may seem clever, or even intelligent, but we suspect that they nevertheless have cognitive blind spots. We have argued that successful psychopaths should be held responsible for what they do (Sifferd and Hirstein 2013), but we nevertheless believe that such psychopaths may have subtle executive deficits. Some of these may include susceptibility to certain fallacies, including the sort of susceptibility to the fallacy of false equivalency described above. Psychopaths may also have mind-reading problems that have larger effects on their cognitive abilities. People who are poor mind readers are going to be poor planners, when the planning requires mind reading. For example, planning to go on vacation with someone requires that you plan in a way that is sensitive to that person's preferences, needs, and desires; otherwise, it is going to be a long and painful trip.

Conclusion

We think that the theory of executive processes makes a good candidate for a concrete and measurable locus of human responsibility. Since it is

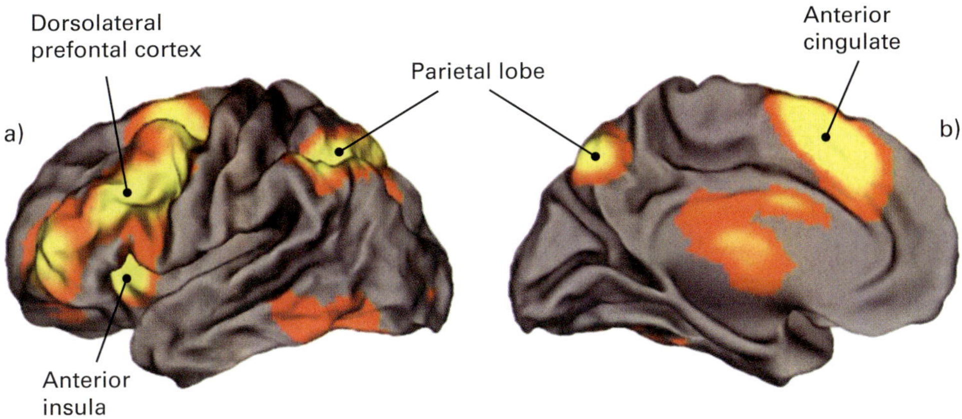

Plate 1

Satterthwaite et al. diagram summarizing findings on adolescents performing a working memory task, the n-back task, which "robustly recruited the executive network" (Satterthwaite et al. 2013, 16253). Subjects are shown stimuli one at a time and must decide if the current one is the same as the one, e.g., 2-back. (a) Left hemisphere, lateral view. (b) Left hemisphere, medial view. Adapted from figure 4, p. 16253, of Satterthwaite et al. (2013).

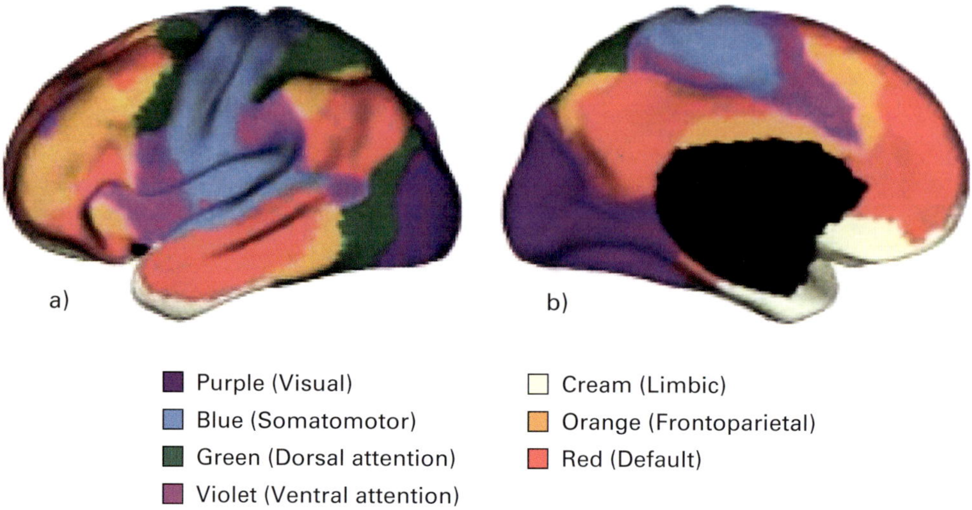

Plate 2

Summary map of Yeo et al., showing the extent of their seven cortical networks. The cognitive control network is in orange and is referred to here as the frontoparietal network. (a) Left hemisphere, lateral view. (b) Left hemisphere, medial view. Adapted from figure 11, p. 1137, of Yeo et al. (2011).

supported by experimental findings in multiple disciplines, and since it has great explanatory power, we think there exists a legitimate *theory* of executive function. The theory has also made itself specific enough to be refuted, should a competitor arise. The discovery of a fairly unified brain basis for the executive processes, the cognitive control network, promises a grounding for the executive theory in brain biology itself.

Many questions remain, however. At this point we will respond to some specific objections that may have come up in the minds of certain readers. We have already dealt with one basic objection, the idea that it is too early in our study of the brain to know whether the concept of executive processes will endure, at the beginning of this chapter. In short, it looks good so far.

One sort of objection we have encountered, primarily from philosophers, is that we are committing something called "the homunculus fallacy." This fallacy is committed when, in the course of explaining something about the mind, a hypothesis makes reference to something like a full-blown human mind. If there is an executive process that is responsible for planning, for instance, should we say that *it* planned my trip? This sounds like the executive process is a homunculus. Normally, the entire human being is the entity that plans. If we speak of an executive process as planning, it sounds like we are explaining human planning by putting a little planner in the brain. As Dennett (1978) noted, however, there is a difference between positing a homunculus while refusing to analyze it further and doing so with the idea that its activities can be given a straightforward physical description. We are beginning to do this with the executive process of planning. For instance, we are beginning to understand how more complex plans are built from simple, routine behaviors, and how the reinforcement history of previous relevant behaviors is taken into account in the planning process.

It is also important to note that the research into executive functions is revealing both their capacities and their limitations. Apparently, incoming perceptual information has to be thoroughly edited and prepared before consumption by the executive processes—the sort of processing that occurs prior to consciousness, such as filling in the visual blind spot. Many of the executive functions can be broken down into primitives that clearly do not require full human mental capacity to perform, such as the stopping or inhibiting of planned behavior when it activates a memory of a similar plan that had a negative outcome, or when a conflict is detected between, for example, two beliefs involved in the planning of the event.

Are you saying, goes another objection, that people's brains, or executive processes, are responsible, rather than the people themselves?[4] Isn't it the whole person only who is responsible? Our title, *Responsible Brains*, might best be thought of as having been arrived at by starting with the concept of a guilty mind—for a person to be responsible, they must possess certain mental capacities and states. Then, on the hypothesis of materialism, one changes "mind" to "brain." As stated above, we believe we are building a bridge by which we can understand the mental capacities necessary to responsibility in terms of their underlying brain states. We are saying, in the title, that the secret to our ability to be responsible traces to an understanding of how the brain's highest-level processes operate. Of course we realize that people are responsible and not brains—we are simply emphasizing that it is our brains that make us responsible.

Given this, our task is to discern the brain processes most essential to making us responsible beings. Clearly, some brain processes are more important than others in this regard. In asking what these processes are, we are like the primitive man who wants to know how these mysterious things called cars operate. What is it that makes them stop and go?, he asks. It serves no purpose to correct him and say, No, *the entire car* stops and goes. He wants to know about the engine and how it makes the car go, and the braking system and how it makes the car stop. We want to know, What in us makes us responsible?

Another objection concerns who exactly is in charge here. Is it the case that when planning needs to be done, goes this objection, I activate the executive process of planning? This would be problematic in that there is something referred to by "I" that causes an executive process to activate. This sounds like a homunculus. This is not our view, and we will try to avoid all such ways of speaking although sometimes doing so is difficult. Our view is that the activation of the executive process of planning *is* my planning, and that no such separate "I" is needed. To take another example, I do not voluntarily control the executive processes. Rather, their work constitutes (a type of) voluntary control. If the goal is a completely physical description of the mind, then in general the words of that description should refer to physical things and properties. This includes the word "I." More on this in chapter 6.

In the next chapter, we will describe how executive processes play a role in making humans responsible. We will also attempt to sort out some of the specific roles played by the different executive functions in making us responsible. In addition, we will introduce the concept of a *minimal working set* of executive functions, that is, enough of the executive functions operating at a high enough level to enable a person to engage in the cognition needed to be a responsible agent.

3 Moral Responsibility

In this chapter we will describe some of the core practices that give rise to responsibility. We will then examine two popular theories of responsibility: reasons-responsive or control-based theories, and real-self (or deep-self) theories. We next introduce our executive theory of responsibility, which is more closely aligned with reasons-responsive accounts.

Folk Concepts of Responsibility

The case studies discussed in chapter 1 involve criminal defendants. But, of course, holding persons responsible is a much larger phenomenon than assessing criminal guilt. Before moving on to an examination of the folk concepts underpinning criminal responsibility in the next chapter, we will consider the larger issue of moral responsibility.

"Responsibility" is a term that picks out a cluster of overlapping folk notions. These notions include (among other things) virtue responsibility, the responsibility an agent has for possessing certain character traits; role responsibility, responsibility based upon the duties one has given one's role in the circumstances; and outcome responsibility, for the outcome of a person's actions or omissions (Vincent 2011, 15–19). In a paper that attempts to taxonomize responsibility concepts, Vincent shows how judgments about one kind of responsibility might shape or influence judgments about another kind of responsibility. For example, judgments about capacity responsibility influence judgments about outcome responsibility and the extent to which one may be liable for one's actions—liability responsibility (Vincent 2011). This is because an agent's capacities determine whether it is possible for him or her to obey moral and legal rules. A lack of capacity may then provide an excuse for an agent.

Our primary aim here is to provide an account of how persons are responsible for acts, omissions, and consequences. Thus, on Vincent's (2011) taxonomy we are interested in the moral and legal capacity to perform acts (capacity responsibility), responsibility for the outcomes of action or inaction (outcome responsibility), and whether sanctions may be warranted (liability responsibility). To discuss these, however, we must first explore our natural tendencies to assign blame in response to wrongful acts.

Reactive Attitudes

We agree with P. F. Strawson that our reactive attitudes are a powerful factor behind everyday judgments of moral responsibility. In his famous paper "Freedom and Resentment," Strawson claimed that the attitudes expressed in holding persons morally responsible are generated naturally from our participation in personal relationships. Such attitudes express "…how much we actually mind, how much it matters to us, whether the actions of other people—and particularly some other people—reflect attitudes towards us of good will, affection, or esteem on the one hand or contempt, indifference, or malevolence on the other" (Strawson 1962, 5). We react powerfully to actions that express a person's good will, ill will, or indifference to others. Judgments about the appropriateness of particular responses to behavior thus seem to hinge upon whether a person had the capacity to perform a morally relevant action, and whether the person's behavior was an expression of ill or good will.

Imagine you are sitting facing the window of a café looking out onto a village green. You see an interaction between a group of children. One older, larger child—maybe 14 or 15 years old—points and laughs at a younger, smaller child, who looks unhappy, and then the older child pushes the younger child down and continues to point and laugh. The younger child then begins to cry. In watching this interaction, it is likely that your moral anger will be triggered: You may use your mind-reading capacities to infer that the older, larger child is bullying the younger, smaller one, and this may make you feel angry—maybe even angry enough to go outside and intervene, especially if it looks like the older child will physically hurt the younger child.

Reactive attitudes such as moral anger seem backward-looking in that they are triggered by, and fitting with regard to, the features of some act (in this case, the actions of the older child). Further, the reactive attitude of moral anger justifies and catalyzes the placing of blame: You blame the older child for bullying the younger child, and your moral anger signals his blameworthiness. You might say to yourself, "Wow, that kid is such a jerk! I hope someone steps in and tells him off."

Now imagine you watch the younger kid gather himself up off the ground and push the older bully so that he trips over a rock and falls down. The crowd of children then laughs at the bully. You might think in this case, "Yes! That big kid got what he deserved." Your moral anger was the vehicle by which you blamed the bigger child; falling down and being laughed at by the other children was something the child deserved given his poor behavior. However, if the push from the smaller child had resulted in a severe head injury to the bigger child, you might rush out to help, thinking, "Well, he deserved to be knocked down, but he didn't deserve to get hurt!" Your notion of what the older child deserves is likely to reflect your sense of what is proportional to the act that triggered your moral anger. If the younger child had been seriously hurt when he was pushed down by the older child, we probably would not think that the older child deserves to suffer similar serious injuries; instead, we think in this case it might be proportional to involve his parents or the police so that they can determine what sort of punishment might be appropriate.

Thus, basic desert—the idea that a person may deserve moral praise or blame—seems to be a natural outgrowth of responsibility attributions grounded by reactive attitudes (Strawson 1962). If moral responsibility attributions are fundamentally responses to some action or actor, then one function of such responses is to express praise or blame. Whether a response is fitting depends upon the nature of the action. If you became angry and resentful toward an older teen not for bullying, but for helping a younger teen find the library when he asked for help, your moral anger would not be fitting. People would find your reaction inappropriate, or at least mysterious.

To determine the fit of a reactive attitude, one must look backward in time at the actions that triggered it; again, fit is backward-looking. If you think, seeing the bully, "His behavior is unacceptable and someone should stop him!" you probably do *not* mean, "Someone should do something so that he is less likely to bully in the future." Instead, you sincerely believe he deserves to be humiliated by the person he is bullying—even if he doesn't learn anything from this humiliation. You might hope such humiliation would have an effect on his future behavior, but the fitness of your reactive attitude does not depend on that prospect. Similarly, if your friend forgets his elderly mother's birthday, you might say "That's terrible! She's probably very hurt!" You might also hope that your censure will have the effect of helping him remember next year; but regardless of the chances that this lesson will be learned, it is fitting for you to be disappointed in him. If his mother were to die that evening, it would probably increase your disappointment and his guilt for the lapse, not lessen it.

We are especially interested in cases where our reactive attitudes are influenced by facts about a person's mental states that seem pertinent to his or her misdeed. That such cases exist suggests that our natural reactions take an agent's mind into account in producing the reactive attitudes. As we will see in what follows, not just any mental state or event on the part of the target person evokes these attitudes. What seems to evoke them most strongly are cases where there is evidence that the target person intended or planned an action. As we will argue in this chapter, signs of executive activity can be evidence of both agential capacity and ill or good will (in Strawsonian terms). The more involved the executive is in producing an action, the greater the likelihood a person will be judged to be responsible for that action and its effects. Or so we will argue as we describe our approach.

Reasons-Responsive Accounts of Responsibility

Many philosophers claim that assessing responsibility involves discerning whether the target person possesses certain cognitive capacities. Given the discussion above, this is not surprising. (Indeed, what would be really surprising would be a theory of responsibility that *didn't* rest at least partly on cognitive capacities.) However, the idea of responsibility as dependent on cognitive capacities has caused some to wonder if the capacities in question operate in a way that actually undermines responsibility attributions. Many contemporary philosophers and scientists are physicalists, who believe that all and only one type of stuff exists in the world: material, or physical, stuff. A physicalist worldview generally includes the idea that every event, including human action and decision-making, is either necessitated by antecedent events and conditions, together with the laws of nature, or is indeterminate (thus, random). That disjunction may have worrisome implications for responsibility: If either determinism or indeterminism is true, then human decision-making and action hardly seems to count as "free." Every decision or action can always be traced back to influences outside the agent: to the agent's genetics, to the agent's upbringing, to features of the environment, or to random events. If responsibility requires our decisions be free in this sense, the idea that the world is made up of all and only physical stuff may undermine attributions of responsibility.

The past 40 years have witnessed a rise in the popularity of compatibilism, whose champions largely reject the idea that responsibility hinges either on an agent's ability to do otherwise in a libertarian ("uncaused causer") sense or on an agent's being the initial cause of the action, separate from other antecedent causes. Instead, on compatibilist accounts of responsibility, an agent's responsibility for an action generally reflects facts about the relationship of the agent's mind to his or her action. And those facts depend on whether the agent had the right kinds of mental structures or capacities at the time he or she acted, even if those structures or capacities function in a mechanistic or deterministic way.

One of the most popular compatibilist theories is that of John Martin Fischer and Mark Ravizza, according to which agents must have the mental capacity to be "reasons-responsive" in order to be held responsible for their acts (Fischer and Ravizza 1998). Their theory revolves around the idea that we are morally responsible for an action when the mechanism issuing that action is *moderately* reasons-responsive in an appropriate way. By "mechanism" Fischer and Ravizza mean "the processes … that actually lead to the action" (Fischer and Ravizza 1998, 38). This claim has three components:

1. An agent must be regularly receptive to a range of reasons, including moral reasons, such that he or she manifests an intelligible pattern of responsiveness (over time) from a third-person perspective.
2. To be held responsible for a particular act, it must be shown an agent would react to at least one sufficient reason to do otherwise than the agent did in some possible scenario—however, it does not follow that the agent could have responded differently to the actual reasons.
3. An agent's mechanism is appropriately reasons-responsive only if the agent has come to own that mechanism by means of a process whereby he or she takes responsibility for the mechanism giving rise to his or her actions (Fischer and Ravizza 1998, 207–39).[1] Agents may take responsibility for their reasons-responsive mechanism by viewing it as the source of their behavior, by believing they are apt candidates for

Strawsonian reactive attitudes, and by viewing themselves as responsible agents.

Fischer and Ravizza thus maintain that when an agent takes responsibility for his or her mechanism of action, the agent "owns" that mechanism. When an agent's mechanism is moderately responsive to reasons, he or she acts culpably.

Fischer and Ravizza's reasons-responsive account of responsible agency may help us understand the "could have done otherwise" notion that lurks beneath the libertarian free will—free will in an "uncaused causer" sense. This notion seems appealing to the folk in at least some circumstances (Nichols and Knobe 2007). People tend to show sympathy for the view that persons are responsible if they could have chosen not to perform the harmful act for which they are blamed. A reasons-responsive compatibilist might claim that a responsible agent needn't have the contra-causal ability to do otherwise with regard to any particular act; instead, a responsible agent must have a mechanism that is sensitive to reasons not to perform the harmful act.[2] This sort of claim lies at the heart of many compatibilist views of responsible agency.

The compatibilist project was aided by Frankfurt's (1969) discovery of cases suggesting that people can be held responsible even if they have no alternatives. For example, suppose Bob plans to kill Paul. But unbeknown to Bob, an evil neuroscientist has gained access to Bob's brain in such a way that he can prevent Bob, in case he begins to have second thoughts, from deciding *not* to go through with the murder. The neuroscientist never has to use this "override," though, because Bob goes through with the murder as planned. Frankfurt argued that in this case, we would and should hold Bob responsible, even though he "could not do otherwise" than murdering Paul. And if that is right, then the ability to do otherwise is not necessary to responsible action.

For a reasons-responsiveness theorist, Bob is responsible because he has a sort of control over his behavior, termed "guidance control" by Fischer (1994). As already mentioned, a person exerts guidance control over his or her own actions insofar as they proceed from a moderately reasons-responsive mechanism (Fischer and Ravizza 1998). Again, this just means that the agent meets the criteria listed above: The mechanism issuing behavior is the agent's "own" (the agent takes responsibility for the mechanism); the agent would respond to at least one sufficient reason to do otherwise in some possible scenario; and the person's mechanism is regularly receptive to reasons, some of which are moral reasons, meaning the agent exhibits an understandable pattern of reasons recognition. None of this requires the sort of libertarian control that involves the ability to act differently in the actual circumstances (which would require that alternative possibilities be open to the agent).

Some persons, though, lack guidance control over their actions and thus are not fully responsible for them. Individuals with a serious mental illness, for example, might be excused from culpability because their action-producing mechanisms are not appropriately sensitive to reasons for action or because they lack the volitional control to act in accordance with such reasons. A schizophrenic with a persistent and salient hallucination or delusion that God ordered her to act a certain way may be unable to recognize important moral reasons for action—or, even if she recognizes them, may not be able to conform her behavior to these reasons. A 13-year-old is not sensitive to the same range of reasons for action as an adult and will be acutely sensitive to some poor reasons for action (such as the esteem of his or her peers). Further, the impulsivity of juveniles is well-documented and indicates that they have less volitional control than adults (Steinberg and Scott 2003). These topics will be discussed at length in chapters 8 and 9.

Diachronic Agency and Reasons-Responsiveness

Below we will endorse a theory of moral responsibility congruous with Fischer and Ravizza's reasons-responsive model. But first, we will discuss an interesting implication and extension of the model. We feel that Fischer and Ravizza's requirement that an agent must "own" his or her action-producing mechanism, in addition to the requirement that an agent exhibit a pattern of sensitivity to reasons, shows moral responsibility to be historical, or diachronic in nature. The ownership requirement means an agent's responsibility for an action or its consequences can depend on how his or her mechanism came to operate as it does. An agent is likely to take ownership over a reasons-responsive mechanism cultivated over time by its possessor, but bears a different relation to a mechanism with a different type of history—for example, one interfered with by a malevolent neuroscientist. Thus, in this latter case actions issued by the mechanism may not be the responsibility of the agent.

Further, as we will argue below, an agent's active exercise of his or her reasons-responsive mechanism over time allows the agent an important type of control over his or her agency—namely, the ability to manipulate his or her behavior by (directly or indirectly) manipulating his or her responsiveness or sensitivity to certain reasons. Fischer and Ravizza's theory is compatible with work by other philosophers who focus on the development and maintenance over time of the mechanism that allows for responsible action. In this way, an intelligible pattern of responsiveness over time not only indicates a mechanism that is moderately reasons-responsive; it also often indicates something about the agency and values of the agent by revealing the dispositions an agent has developed to be sensitive to certain types of reasons (and not to others).

Aristotle was perhaps the first Western philosopher to emphasize the importance of diachronic self-development and authorship of action to responsibility, and there has been a recent resurgence of interest in the topic among contemporary philosophers.[3] In the *Nicomachean Ethics*, Aristotle (1985) argues that agents are responsible for habituating their characters, conceived of as dispositions to act (1105b25–26). Habituation involves practicing the trait via the use of practical reason, which allows a person to determine which actions are appropriate in a given situation. A person builds a virtuous character trait, such as honesty or courage, by making honest or courageous choices over time and across situations, resulting in a stable disposition to act honestly or courageously.

From the perspective of a reasons-responsive account of responsibility, habituation may be a way an agent can alter the mechanism that issues behavior for which he or she can be held responsible. We may view a reasons-responsive mechanism's sensitivity to certain types of reasons as akin to a person's disposition to act in certain circumstances. Imagine Todd decides he wants to be more sensitive to his partner's needs. As a result, Todd sets a reminder to pay closer attention to her schedule and emotional states; and he makes an effort to take on some of her responsibilities when she is very busy and to comfort her when she is anxious. As time passes, Todd no longer has to consciously remind himself to pay attention to those things in order to be sensitive to his partner. He has habituated such sensitivity, and he doesn't have to deliberate regarding whether she is anxious or needs comfort—he has developed a feel for it. What Todd has done, on the view we endorse, is manipulate his reasons-responsive mechanism so that it is more easily and quickly responsive to certain reasons for behavior—namely, his partner's needs. And, when we view Todd's pattern of responsiveness over time, we see him forming a disposition to be sensitive to certain reasons.

Adina Roskies (2012) has argued that this sort of diachronic rational self-review and intervention can ground a compatibilist account of responsibility that preserves real agency but avoids the "obscure metaphysics" of libertarianism (323). We can deliberately intervene in our future selves, says Roskies, by manipulating our mental content in ways that have foreseeable consequences, and because we have such diachronic control we are responsible not only for what we do, but for who we are. Roskies has an expansive sense of the possible forms that diachronic self-intervention can take: An agent has the ability to engineer his or her environment so that it elicits or makes manifest valued dispositions, and does not realize disvalued ones; to intervene on his or her future self by making commitments to future behavior or setting overarching policies; and to strengthen the processes of self-control by, for example, practicing decision-making in a way that increases deliberative control (Roskies 2012). Alcoholics can avoid walking past their favorite bar on the way home from work; individuals who wish to exercise can join a yoga class with a friend who will pressure them to go; and people can imagine their grandmother's disappointed face every time they consider not going for their weekly visit. In these sorts of ways persons may not just directly manipulate the operations of their reasons-responsive mechanism by establishing dispositions; they may also encourage or discourage certain decisions by exposing their mechanism to certain environments and not others, and enhance their sensitivity to certain reasons by framing decisions in a certain way.

Like Roskies, we believe that the diachronic exercise of top-down higher-level reasoning processes to manipulate one's decisions and actions is important to responsibility and agency. Deliberation and choice can have downstream effects—according to the reasons-responsive model, they can manipulate the mechanism issuing responsible action. And these effects may result in actions that are habitual or done without conscious awareness, but for which an agent may still be responsible. Our theory thus designates a larger range of acts for which we can be held responsible than would be suggested by a purely synchronic assessment of the cognitive capacities engaged when an action is produced.

We will argue that executive functions are the primary means by which we can habituate a disposition to act and can strengthen self-control by manipulation of self and environment, and thus executive functions are the means by which we can intervene in our future selves. If agents can consciously deploy executive processes to manipulate their own dispositions and future choices, such agents may be responsible for actions stemming from such dispositions and choices, even if those actions are not related to synchronic conscious states representing ill or good will. We will expand our argument for this position in chapter 7, in response to arguments by Neil Levy that conscious awareness of the moral implications of one's act is necessary to responsibility. And in chapter 10, we will discuss how different criminal punishments can enhance—or undermine—such diachronic agency.

Real-Self Theories of Responsibility

In this section we will compare the reasons-responsive account of responsibility to its primary competitor, the real-self theory of responsibility. Whereas reasons-responsive views claim agents must in some sense have control of their actions via the capacity to be reasons-responsive, real-self theories claim that an agent must have a characterological or expressive connection between certain psychological structures central to the self and action. The real-self camp arguably originated with Harry Frankfurt (1982). On Frankfurt's theory, an agent acts freely, and thus potentially bears culpability, when he or she acts in accordance with his or her "deep" or "real" self, which is manifest in higher-order values or desires. Moral responsibility thus depends upon whether or not an agent reveals his or her real self in acting as he or she does. A person who acts in accordance with a first-order desire—to, say, eat a piece of cake—is responsible for that action if (and only if) it would be reviewed and endorsed by that person's higher-order desires or values. However, if a first-order desire is not in agreement with second-order values—for instance, because the first-order desire is so overwhelming or salient that it is unreviewable—then the act related to it may not reflect the agent's real self. For example, if the act of eating the cake is related to the overwhelming first-order desire of someone with an eating disorder or a food addiction, and the agent disavows the action because it conflicts with his second-order desire to be heathy ("I don't know what came over me; I wasn't myself"), then the agent ought not be held responsible for eating the piece of cake. On Frankfurt's view, in this case the cake-eater is an "unwilling" addict.

On Fischer and Ravizza's reasons-responsive view, the unwilling addict may also qualify as exempt from responsibility. That exemption would not depend on a mismatch between his action and his higher-order desires, however, but on the overwhelming nature of his desire for the cake, which caused him to fail to manifest an intelligible pattern of reasons-responsive behavior over time. Alternatively, the addict might be exempt from responsibility because he lacks the capacity to react to at least one sufficient reason to do otherwise when he eats the cake. In the specific context of addiction, Chandra Sripada puts the difference between reasons-responsiveness and real-self theories this way:

> *Reasons-responsiveness* views focus on the addict's control over his actions. Very roughly, to be morally responsible for an action, the person must be able to, across a suitably broad range of scenarios, issue the action when there is sufficient reason to do so, and withhold the action when there is sufficient reason not to. *Deep* [or real] *self* views, in contrast, say the addict is not responsible for using the drug because his action does not express his *self*. That is, since the addict's own fundamental values and principles repudiate the action, the action does not reflect the person who he truly is, and he is not responsible for it. (Sripada 2015, 242)

The two views come further apart in the case of the "willing" addict—an addict who endorses or values his addiction. This person would still seem to lack the control required for responsibility under a reasons-responsiveness view but may be held responsible on a real-self account because the addictive behavior *does* express the addict's real self.

Our theory aligns more closely with a reasons-responsive account of responsibility, for a few reasons. First, the capacity for synchronic and diachronic rational control of behavior is extremely important to responsibility. For example, a person who has the capacity to pause and inhibit a harmful act, such losing his temper and punching another person, may be held responsible for that action even if it is out of character, or not endorsed by the person's values. We cannot easily determine, in such a case, whether the urge to hit the other person was so overwhelming that it could not be reviewed by second-order desires, and at any rate this is not the right question to ask. A better line of inquiry focuses on whether the person had the capacities to review the decision and inhibit it—in Fischer's and Ravizza's terms, whether the person was moderately reasons-responsive. Our theory claims that the sorts of rational capacities manifest in a reasons-responsive mechanism should be understood in terms of the brain's executive processes. In our terms, if the person had a minimal working set of executive functions (MWS), and if the person's executive functions played the appropriate role in generating the action—or should have done so—that person is responsible. The idea of an MWS gives a new and better way to specify the capacities Fischer and Ravizza gesture at when they claim a mechanism must be moderately reasons-responsive to issue action for which we can be held responsible.

Having said this, we suspect the two perspectives may be closer in their account of responsibility than it initially seems. Evidence that an act is out of character may sometimes indicate that a person was not reasons-responsive when he or she performed a criminal act. Imagine a mother who is generally thought to be a kind, compassionate, and responsible person who leaves a five-month-old infant alone in her apartment for three hours while she goes to the gym. We might suspect that the anomalous nature of the act indicates she is suffering from postpartum depression. If that suspicion bears out, it would not be the act's failure to express her character which might excuse her from responsibility—it would be her failure to meet conditions of reasons-responsiveness. However, the act's lack of relation to her deep self is epistemically relevant, because we can recognize that people often do out-of-character things (for which they may not be responsible) when they are suffering from mental illnesses or conditions undermining their reasons-responsiveness.

Noting the shared ground between these two ostensibly opposed theories, Manuel Vargas claims, "If the manner in which one responds to reasons just is a way of expressing one's character, commitments, or values, then the difference between the two views evaporates. Similarly, if one's character, commitments, and values say something about what the agent regards as reasons-giving, these approaches may come to the same thing" (Vargas 2013, 138). Still, even if the rift between reasons-responsiveness and deep-self views is smaller than typically recognized, Vargas argues that the reasons-responsiveness view is preferable for two reasons: First, the strengths of the deep-self view seem to more easily collapse into the reasons-responsive view than the other way around. And second, a focus on the canonical cases of responsible action—exactly the type of case we think criminal verdicts capture—indicate that control matters more than character.

We agree. A reasons account's focus on control also seems to better reflect the folk compatibilist notion that persons are responsible if they had the capacities to have done otherwise than to commit the harmful act for which they are blamed. In his book *Building Better Beings*, Vargas (2013) notes that blame attributions seem to rest upon the notion that the person blamed had a reason to act otherwise than he did. This sense that the blamed agent ought to have paid attention to other reasons relevant to his or her action is obviously more compatible with a reasons-responsiveness account than an account that focuses on the deep self (Vargas 2013). Further, blame can be understood as calling another to account for his or her action: When we blame someone, we are in some sense asking an agent to provide his or her reasons for action (Shoemaker 2015). This process of calling an agent to account may then involve countering the agent's reasons with alternate reasons to do otherwise. As it plays out, this process resembles a collaborative or public process of reasons evaluation, which again, seems to cohere more closely with a reasons-responsiveness account (Vargas 2013, 139). In sum, it isn't because we possess second-order reasons or values that humans are uniquely responsible actors; instead, what matters is the way we may utilize these second-order values and long-term plans as part of a rational decision process to produce complex behavior.

Vargas (2013) also notes that real-self views have a comparatively hard time dealing with cases of a malevolent manipulator. On a deep-self theory in the style of Harry Frankfurt, responsibility would seem to accrue to one acting in accordance with second-order desires regardless of whether they were cultivated normally or put there by brainwashing or implantation—which may well strike us as counterintuitive. The historical requirement of a reasons-responsiveness account, by contrast, seems to handle this problem, at least in some cases: The manipulated agent may not see himself or herself as the source of behavior related to an implanted reason as required by Fischer and Ravizza's model, so the agent may not take responsibility for the behavior. Similarly, on Frankfurt's view, an agent would still be considered fully free and responsible for actions endorsed by second-order beliefs or values that are insane or delusional (Vargas 2013, 143). A reasons-responsiveness account, however, can provide at least some normative guidance regarding the reasons to which an agent ought to be sensitive in order to qualify as responsible.

The Executive Theory of Responsibility

We can now begin building that bridge we promised in chapter 1, connecting existing concepts of responsibility to the brain. When somebody acts—or fails to act—in a way that bears moral import, that person either is or isn't responsible for that act, for that omission, or for certain consequences of that act or omission. Our account specifies the mental requirements for moral responsibility in terms of physical, measurable brain states and capacities. In this section we will delineate some of the basic features and future directions of the executive theory of responsibility.

Here is an abbreviated version of our hypothesis, with details, caveats, and qualifications following in separate subsections for each of the three criteria:

Jo is responsible for an act, omission, or consequence only if

1. Jo has a minimal working set of executive functions, and
2. Jo performed the act, and/or caused the consequence, or failed to act to prevent it, and
3. Jo's executive processes either played an appropriate role in bringing about the action, omission, or consequence or should have played an appropriate role in preventing it.

We do not believe these three conditions together are sufficient since there are other factors, including situational and cultural ones, which may partly determine responsibility. But, we will argue, each of these three conditions is necessary for holding a person responsible for something that person did or caused. Criteria 1 and 3 are directed at the mental requirements for responsibility. Criterion 1 emphasizes an agent's basic capacity for reasons-responsive thought and action. Criterion 3 emphasizes the way that our judgments of responsibility track the presence or absence of executive activity. Its presence helps make a person guilty in the positive cases, and its absence helps make a person guilty in the negative cases, cases in which someone failed to employ his or her executive capacities, but should have. While criterion 3 focuses on "internal" mental factors that are required for responsibility, criterion 2 focuses on "external" factors involving causal interactions between bodies and other bodies and things. Having two criteria of this sort is not new. Criminal responsibility, for example, requires a defendant be found to have had a "guilty mind"—mens rea—and to have performed an act that violates the law—actus reus.

Criterion 1. The Competence Criterion: Possession of a Minimal Working Set of Executive Processes

Executive abilities come in degrees. One person might be a master at planning while another is terrible at it. We each have our executive profile, as it were, which specifies each of our executive capacities and their degree of proficiency. These capacities need to function at certain levels before a person qualifies as competent. However, since executive processes need to interact to achieve many mental functions, and all the executive processes, or at least most of them, must be able to operate to some minimal degree, it seems that one needs a basic working set of them. Each of them—planning, monitoring, inhibiting, forming intentions, and so on—plays a role in making us responsible beings. It also appears, though, that some of them, such as attention, planning, and inhibition, have dominant roles among the mental components required for responsibility.

We might draw an analogy with our use of cars in order to illustrate the idea of an MWS. In the interest of public safety, cars must meet certain standards before they can be driven on public roads. A minimal working set of functions that a car must have includes acceleration; braking; turning; perceptual aids for the driver such as headlights, windshield wipers, and rearview mirrors; and the capacity to communicate to other drivers, including headlights, running/parking lights, turn signals, and brake lights. Notice that some members of the minimal working set are "social" in that they exist both for the driver and for those the driver encounters on the road, such as the car's system of lights. Similarly, some of the executive functions have social functions built into them, for example, planning for actions that will involve others, simulating their responses, and then possibly inhibiting a planned or intended action based on an educated guess at the person's response. There are additional automotive functions that make life easier and safer for the driver but which are a bonus as far as street legality is concerned: cruise control, power steering, antilock brakes, a powerful engine, fog lights, and so on. Similarly, most people have a set of executive functions that far exceeds the minimum that allows one to participate in a society.

There are apparently many different ways to achieve a minimal working set—several executive styles, as it were. Some play a prominent role in our folk conception of personality. Consider the highly inhibited person, contrasted with a person who is wild and expressive, but within socially acceptable bounds. People who make a habit of lying to others, to use another example, are certainly employing their executive processes in different ways from people who tell the truth. Lying successfully can be quite complicated from an executive point of view, requiring planning, top-down mind reading (to make the lie convincing to the listener), careful alterations to memories, the addition of realistic details, and so on. Powerful social forces are also at work normalizing our executive processes and their amalgamated profiles, especially as we grow up. Parents, peer groups, schools, media, and sometimes the justice system itself ensure that we make efforts to strengthen executive functions that are weak. It is easiest to see these forces at work on a child with weak inhibitory functions, but they also work on those with weak planning capacities or those who are weak at considering and correcting initial perceptions.

Each of these executive profiles will respond to and recover from deficits or damage in a characteristic way. If other executive functions are intact, one can at least partially correct for a memory deficit, for example: Persons can use aids, such as notes. They can be aware that they ought not trust certain memories, and so on. Theory of mind deficits can also be partially corrected for: Some people with autism spectrum disorder, for instance, memorize scripts appropriate to social interactions they are preparing to engage in. This helps them interact with others without being able to understand their minds in the natural and spontaneous way that many people can.

Sometimes brain damage will temporarily disable a person's executive processes severely enough to prevent that person from participating in society in the normal way. Correcting for the loss of an executive function caused by brain damage in adulthood can be quite difficult. Certain patients with frontal damage, for example, have lost the ability to inhibit certain types of behaviors and display a disorder known as perseveration: Even when they are aware that they should inhibit a response, they simply cannot do it. Such a patient might be instructed by his neurologist to sort a deck of cards by suit, separating diamonds, spades, hearts, and clubs. He is able to do so, but when told to instead begin sorting the cards by color, he is unable to stop sorting them by suit. He knows that he is failing to follow the command and can even verbally express that. He wants to follow the command, but he cannot stop himself.

There are also serious executive deficits that are successfully overcome, however, both during the developmental periods and following stroke or traumatic brain injury. Here is one example: An apparent mark of attention-deficit/hyperactivity disorder (ADHD) is that the anterior cingulate, an important part of the cognitive control network that performs high-level executive functions, is not used to switch between cortical networks as the task demands, as in the normal case (Zhan et al. 2017, Bauer et al. 2016). Yet people with ADHD in general function quite satisfactorily in society by using other, somewhat less effective, brain processes to switch their attentional focus. Obsessive-compulsive disorder (OCD) may be another case where people compensate for a malfunctioning executive process, an important inhibitory one, that is hyperfunctioning, stopping all thought and action until, for example, hands are washed or door locks are checked yet again. In this case, people can achieve the minimal working set needed to participate in society by using different therapeutic techniques, medications, or even surgery in extreme cases.

Thus, the different executive functions have some ability to make up for each other's deficits, and this is important for our approach to responsibility. It means that one isn't permanently stuck with the set of executive processes one has now, doomed to forever suffer from their weaknesses. We can devise strategies to correct for our deficits and train ourselves to notice and take action when they might cause us to do something unethical or illegal. To continue with the car analogy, these corrected executive deficits would be like a car with malfunctioning brakes which is still able to function on the roadways by use of the emergency brake instead. It doesn't work as well, and there is no backup for it, but it works well enough to allow the car to be driven. Drivers in this situation must alter their driving behavior to correct for it. They do not have much braking power, so they need to take their foot off the accelerator sooner than normal and begin to apply the brakes sooner. The ADHD sufferers we described above appear to be doing this by performing an executive function using a nonnormal set of cortical areas. And some people with OCD are able to functionally isolate a hyperactive executive process, an inhibitory one, from having widespread effects on their cognition and mental lives.

But surely, even though the executive processes need to work together, they are not all equally important to our abilities to be responsible members of a society. Which executive processes, then, are the most important for responsibility? Are any of them dispensable? At the outset, it is important to keep in mind that executive deficits do not parse neatly into cases in which one or more executive processes are completely absent.[4] The deficit can be a matter of degree, or it can be caused not by damage to an executive area, but by damage to an area that the cognitive control network normally interacts with as part of its functioning, a list that is large and diverse, including perceptual areas, motor areas, mnemonic areas, and the default mode network, as well as areas involved with emotion.[5] Another thing to bear in mind when reviewing the scientific literature on patients with executive damage is that the clinical examples of loss of a single executive process may be odd or exaggerated, partly because those people either have not had time to recover or have been unable to and, hence, are institutionalized and available to researchers. Further, within this clinical population there is a selection bias in that neurologists and neuroscientists specifically look for patients who have brain damage that is as focal as possible because this makes the task of localizing the relevant functions easier.[6]

Planning is obviously an important executive function with regard to responsibility. Since planning must depend on more than one representational system, it should be possible for a person to lose the ability to form a certain type of plan while retaining the ability to form one or more other types. Deficits in planning should show a pattern of specificity determined by the adjunctive representational system the cognitive control network interacts with to accomplish that type of planning. For example, if I need the default mode network to assess how desirable a possible future state of affairs is in order to plan, say, a robbery, and the network is not functioning, my plans will be poor or nonexistent. Inhibitions that arise during planning play a vital role in the process—I use inhibitions to correct and edit the plan—which suggests that each type of planning may have its own variety of inhibition. A loss of an ability to plan will have negative impacts on one's ability to carry out a certain class of complex, future-directed actions. Sometimes a person can enlist the help of another person to overcome executive deficits. Someone who is poor at planning may be prudent enough to marry someone who plans quite well. But people with serious enough planning deficits sometimes have to live at home with parents or other caretakers or in institutions.

As we noted, there appear to be different varieties of inhibition, and some are clearly required in order to function in society. One need only look at the neurological literature on patients who have become *disinhibited* (Starkstein and Robinson 1997). They commit acts that violate moral norms as well as laws and can develop social problems, such as becoming rude or inconsiderate. Inhibition itself is a mere stopping force. It can stop a contemplated action from becoming a planned action or from becoming an intended action. But some actions need to be stopped for a good reason. This suggests that inhibitory mechanisms need to functionally interact with high-level systems for representing behaviors. Such representation systems are also likely to have connections to the brain's emotional and reward systems, which attach emotional and/or reward valences to the representations. For example, in trying to decide where I will go to eat this evening, I imagine myself in three different places. Each representation is accompanied by an emotional response that I use as a guide in making the decision. And I inhibit an impulse to go to a place where the food was poor last time, when I become aware of that negative value tag.

Disorders of attention also take characteristic forms. Loss of certain types of monitoring processes will cause a reduction in the quality of the person's perceptions and beliefs, but sometimes the person's basic, pre-executive perceptual or mnemonic capacities are still sufficient to allow the person to function in society. For example, in the realm of attention to memories, someone might have a quite accurate memory but be unable to use executive processes to improve that accuracy level by attending to and eliminating inconsistencies in apparent memories, checking them against other sources, and so on. In this case, damage to that person's basic memory systems can be disastrous and lead to confabulations since the person never developed the ability to use executive processes to check and correct apparent memories (see Hirstein 2005). Attention seems crucial to staying on task and following a plan, and may be important to correcting for perceptual deficits.

In contemporary society, we often arrive at collective decisions by a process of discussion or debate among opposing voices, ideally allowing the group to choose the best course of action among competitors. Our individual decisions may also be reached via a sort of debate occurring in our brains between opposing executive processes. Certain executive processes are naturally opposed to one another, the most obvious example being that of intention formation and inhibition. Inhibition can oppose planning, or intention, just as the brake pedal opposes the accelerator pedal. Plans themselves are also inhibited, prevented from forming into intentions, when the planning process reveals possible outcomes that are undesirable. As we saw in chapter 2, the executive system is set up to activate when conflicts of several different types are detected and then initiate processes that work to resolve the conflict. For example, in planning medical appointments, conflicts in dates are typically encountered that need to be worked around. This requires task switching, controlled memory recall, and attention to the relevant representations. Similarly, some of the brain's mind-reading systems can alert when a simulation of a person's mind contradicts other knowledge we have about the target person. The cognitive system is set up so that conflicts make it dig deeper, using executive processes, and initiate the gathering of more information, whether via perception or by searching one's memory. These conflicts need to be "honored" by the executive processes and not merely denied or ignored, as happens in one type of self-deception.

Our chosen term—*minimal working set*—denotes the idea that a person needs a certain level of executive function to be the subject of responsibility attributions, and that level of total function must be achieved by the executive functions working together in an effective way. *Minimal working set* is ultimately a functional notion: It compiles a set of functions the person's brain must achieve in order for that person to effectively function in society. (As we shall see in the next chapter, the notion has interesting parallels with legal scholar H.L.A. Hart's notion of capacity responsibility.)

Finally, we claim that when we, in our normal social interactions, assess someone as being fit for assignments of responsibility, we assume that person has been in possession of a minimal working set of executive processes for some time preceding the action in question. This nature of this assumption will become clear as we proceed, especially when we look at cases involving spontaneous behaviors that erupt without any evidence of executive processing proximal to the event. These cases need to be treated diachronically: Even if the person did not use his or her executive processes at the moment of action, it may be true that the person *should have* used them, over an earlier stretch of time, to shape his or her proneness to spontaneous acts of that sort.

Criterion 2. The Causal Criterion: Performance of an Act, or Failure to Act

Human behavior and mental events leading to behavior do not easily come apart in our normal understanding of one another. We do not see people as simply physical objects crashing into one another; we are powerfully disposed to see human interactions as mind driven. In the paradigmatic case, where we have a clear view of someone performing an action, we get an initial take on intent and voluntariness just by using our normal vision. We observe how the person performs the action, especially what the person attends to. When it is important to do so, we also use mind-reading capacities to increase the accuracy of that initial take. This suggests that there is overlap between our epistemic grasp of the external causal components and the mental or internal executive components behind our assessments. For example, the brain has perceptual areas that specialize in perceiving "biological motion," that is, motions of living creatures (see, e.g., Rizzolatti et al. 1996). Using systems like this, we suspect that even without the use of mind-reading capacities, our body-observing capacities would still make basic distinctions between intentional and nonintentional actions by tracking behavior alone. However, the addition of our mind-reading systems increases the accuracy level of these judgments by taking additional information into account. For example, if I know about someone's mind and character, this can help me conceptualize an act as a playful nudge rather than a malicious shove, even if I've never seen that person perform such an action before.

Criterion 2 thus refers to human actions, along with some minimal mental component (which is inevitably present if criterion 1 is met). Criterion 2 also mentions the failure to prevent a consequence as a requirement for responsibility for omissions or their consequences. The notion of preventing need not be mysterious: One can prevent x by intervening in the causal chains that would lead to x. Preventing is a doing, not a nondoing—and prevention is one of the tasks served by the executive processes. Making and following a plan, for example, is one way to prevent undesirable outcomes: One plans a trip to prevent getting stuck at a train station or plans a bank robbery to prevent apprehension and arrest. Using the cognitive control network coupled with the default mode network, we undertake a simulation of future events and then edit the plan or abandon it altogether when the simulation indicates undesirable outcomes. We hold people morally responsible not only for the negative consequences their actions cause, but for negative consequences they reasonably could have foreseen and acted to prevent. Criterion 2 encompasses both sorts of circumstances.

Criterion 3. The Executive Functions Criterion: Is Executive Activity Present, or Should It Have Been?

In many "positive" cases of responsibility—those in which a person's responsibility is for something he or she did rather than failed to do, physical events causing the actions specified under criterion 2 must involve the executive functions in the right way. In most cases, one or more executive processes must participate in the causal chains that bring about the relevant actions. In other words, they must play a role in the production of that action.[7] These causal chains, and the causal interactions between the executive processes and representations of all sorts, can satisfy some of the core mental requirements for responsibility. In these cases, the third criterion specifies that the positive acts specified under criterion 2 (as opposed to, e.g., nondeliberate omissions) must be performed intentionally, with the right sort of executive involvement in their causal history.

We conceive of *roles* simply as (types of) causes within a functional system. Functional systems are made up of roles and occupants that fill those roles. In the circulatory system, the heart occupies the role of pumping blood at the right rate and pressure, for example. The executive function of planning (its occupant: the cognitive control network) causes the creation of a plan, which itself will play a certain role in any action that results. The idea of a process playing a role in the production of something seems straightforward. Some milk is pasteurized and some isn't, the difference being whether the process of pasteurization played a role in the production of that milk. Some actions are planned, some aren't. The difference is whether the executive process of planning played a role. In a positive case, planning might play a role in the production of the action of traveling to Africa. In failure cases, for instance a failure to prevent a bad vacation by planning better, planning is one role that, had it been filled or performed, stood a good chance of preventing the disaster. That person should have planned better.[8]

The same outward behavior can have different executive processes playing different causal roles in its history, and these differences can result in dramatically different levels of responsibility. For example, a man driving in Manhattan accelerates his car into a crosswalk full of people, injuring several of them. One causal history of that event contains executive processes playing planning roles, setting goals to harm as many as possible, and then forming and executing the final intention to accelerate into the people. Other causal histories might involve other executive processes, or none at all. For instance, in another case, the man was distracted by other thoughts and was driving an unfamiliar car, and he mistakenly pressed the accelerator pedal instead of the brake pedal. This particular history contains executive processes, but in a much less culpable way than the first one. In another case, the man had a seizure, so there was no relevant synchronic executive activity. Here, we need to know more about the man's history of seizures. If we find that he has had several seizures that caused accidents, we will consider him responsible for failing to use his executive processes to play roles to prevent future accidents—for instance, by giving up his driver's license.

Other things being equal, when executive processes play roles in the causing of an action, they tend to make that action more culpable. Indeed, more executive activity seems to correlate with greater culpability. For example, planning an action (along with other mental acts) is a way of causing an action. As we noted, not all actions have planning in their causal histories, their etiologies. Those actions that do involve planning tend to be more culpable, but there are culpable actions without any planning role in their history—for instance, the actions of a man who has learned martial arts and kills someone in a fit of rage. People can still be held responsible for spontaneous actions such as this because they involved executive functions playing (or failing to play) other roles, such as diachronic roles like monitoring how one is learning the martial arts and inhibiting oneself from learning to use them in a reckless way.

In one type of paradigmatic case, if you create and manipulate adequate conscious representations of yourself harming some person, x, and use those representations to form plans to harm x, and you then monitor your behavior and perceptions as you harm x (to ensure it goes the way you planned), the mental states clause is satisfied. However, there are many ways that executive activity can cause the relevant actions and satisfy the third criterion. We can enumerate several clear cases of relevant executive interactions with the crucial representations, in this case of a target person, x: *Intending* to harm x; *Planning* to harm x in order to achieve y; *Top-down goal formation*, to harm x; *Top-down mind reading* (especially directed at x) toward the goal of harming x; *Top-down attention* to perception in order to detect and then harm x; *Top-down attention* to memories (especially of x) in order to commit a crime against x; *Task switching* in order to plan and accomplish harm to x. In general, the planning of illegal or unethical acts will involve several types of executive activity.

In other cases, a person may bear less responsibility even if her executive functions are involved in causing the act because the person did not have an adequate representation of something or someone. If a man is reasonably convinced he sees his partner (from the back) in a place she ought to be and then embraces her from behind, he is not responsible when he actually embraces someone who looked very like his partner and was wearing the same coat. When a hunter shoots another hunter thinking he is a bear—it was dark and rainy, and the hunter who was shot did not wear orange hunting clothes—we are naturally inclined to treat him as less blameworthy than if he had knowingly shot the other man (assuming his mistake was reasonable). Our mistaken hunter may deserve some blame if he should have waited until his target was in clear view before firing or if he should have thought about the potential of this scenario prior to going hunting and taken precautions against it (e.g., asking his hunting partner to wear orange). On the other hand, if the victim is a member of an animal rights organization unwisely wearing a very realistic bear costume, the hunter may not be morally responsible at all. In this case responsibility seems to shift from the hunter, who could not have been expected to discriminate the activist from a real bear, onto the shoulders of the activist, who reasonably could have prevented the unfortunate event.

Positive Cases: Cases Where a Person Is Responsible because Executive Functions Were Engaged In positive, synchronic cases, a person is responsible if one or more executive processes played the appropriate role in causing the action in the time slice immediately preceding the act. Planned crimes, as well as altruistic behavior, are examples of positive, synchronic cases. Deliberate omissions are also primarily synchronic, in that the person is aware of the need for action in the crucial moment but fails to act (or even inhibits an action). In cases involving responsible actions—doings rather than failures to do something—(synchronically or diachronically) prior to the act there was in the mind/brain of the defendant executive activity sufficient to satisfy criterion 3. Executive activity together with representations of a victim or object, and representations of harm, connected in the right way, can fulfill the executive functions criterion. Attention can play an important role in determining responsibility, as what a person attends to, especially around the time of the crime itself, is typically going to be strongly relevant to the mental states clause. What we attended to and why as we engaged in actions also affects our level of responsibility. Are we attending to our memories of where someone lives, for instance, to go give him a present, or to go murder him? What we attend to is also a product of a type of shaping by us over time. It is in general true that we attend most to the things we value, and over time this fact makes us more likely to attend to certain things out of habit.

In positive diachronic cases, an agent's executive functions played a positive role in producing an action, but over a longer period of time. Actions that result from self-controlled habituation can occur without *synchronic* executive activity, but the agent can still be responsible for the action if executive activities have been diachronically involved in the habituation process. For example, a person might knowingly and deliberately habituate violent reactions to perceived slights, as prisoners are sometimes forced to do for self-protection. If this sort of prisoner reactively punched a prison guard who startled him, he would still be responsible, even if his executive processes did not engage synchronically. (It seems that sometimes when a person is in "fight or flight" mode, the executive functions fail to engage.) In addition, actions committed during flow states (which have deliberate practice and planning in their history) fit our diachronic model. Flow actions, such as those that occur during mindless driving, or during athletics, can result from a habituation process (more on this in chapter 7). Many positive cases will involve both synchronic and diachronic roles for the executive processes, such as the career burglar who has knowingly habituated that practice (diachronic) and is executing his current planned burglary (synchronic).

Negative Cases: Cases Where a Person Is Responsible but Executive Functions Are Not Engaged Negative cases are those in which a person failed to act, but the person is responsible because the executive failed to play roles that reasonably would have prevented the harm. What qualifies as a culpable failure to act will depend on contextual factors, which can affect the degree and type of care or diligence we expect one another to take. Higher stakes demand higher diligence in general, but often standards of care can be quite situation-specific. For instance, suppose Charles is driving a group of children in a school bus he owns as a private contractor. He stops to get gas but fails to notice that a nail has caused a leak in the bus's left front tire, and as a result the tire is seriously underinflated. Later, while Charles is negotiating a sharp right turn on a winding mountain road, the flat tire loses grip and the bus plows off a cliff with great loss of life. Certainly, Charles *could* have checked the tires at the gas station. But we typically don't consider it reasonable to expect people to do this, and our sense of Charles's blameworthiness may be rather light. On the other hand, if his left front tire were found to be completely *bald* in a way that led to the accident, we would hold Charles more responsible: As the owner of the bus, he both could and should have checked the treads on the bus's tires before they reached such a state. Arriving at an assessment of what someone should have done in a context requires specific knowledge of that context.

The appropriate level of salience of items relevant to one's actions is thus a factor. While salience is generally a feature of bottom-up perceptual processing—our ability to directly control which items in our perceptual field are salient is limited—we do expect agents to put themselves in the right perceptual situations in order to perceive important things or events, and we expect them to allocate their top-down attention in appropriate ways. If a stop sign is mostly occluded by a tree branch, it might not be salient to you even if you were paying proper attention, and so you would not be responsible for not attending to it. But if a clearly visible stop sign was not salient to you because you were texting on your cell phone, you are not excused because you did not put yourself in the right position to see it.

The failure portion of criterion 3 also captures the idea that people are often responsible for not remembering important responsibilities, not paying attention to the right things, or not thinking about important consequences of their actions. Synchronic cases here involve situations that require executive activity—the agent needed to pay extra attention, or take extra care, or inhibit an intended action that was immoral or illegal. A man rushing to make an important work meeting who distractedly backs over a first grader in his SUV is responsible for not engaging his executive processes to the extent he should have and checking behind his vehicle. He may be driving using minimally conscious processes; his executive processes are engaged in planning his day when they should have been recruited to drive more safely. Typically, the failure clause is satisfied when causal interactions between executive processes and representations of x (relevant to preventing x) did not happen but reasonably could have and should have, given the context. As with positive cases, negative cases will often involve both synchronic and diachronic executive activity in the explanation of an omission or its consequences. The man who accidentally backs over a child failed to check on that occasion (synchronic) and failed to habituate a checking procedure (diachronic) or undertake the planning required to get a vehicle with a backup camera.

Undermining Responsibility

We do not hold a person responsible for failing to prevent a murder if the person was tied up at the time. And we do not hold people responsible if their normal abilities to plan and intend and inhibit actions are substantially impaired. These can include cases out of science fiction where an evil neuroscientist has used advanced technology to "take over" a person's brain. There look to be at least two broad types of cases that undermine responsibility: external cases, in which the person's body is encumbered or in some way prevented from performing some action, and internal cases, in which the person's decision-making mechanism is hijacked or affected in some way relevant to the action in question. The person's body must be unencumbered and the person's executive processes must be unencumbered—not subject to external or internal pressure not initiated by the agent—in order for a responsible act to take place.

Note that our theory states that *that person's* executive processes, functioning in their normal way, must play the right role in bringing an action about (or should have done so)—not some special laser ray built by an evil scientist or some brain parasite implanted by aliens. This rules out the possibility that any of these deviant cases could work as counterexamples to our theory by fulfilling its conditions in cases where the person is not responsible. In fact, these examples work to affirm our theory by honoring the idea that something in our brains could be the special thing that makes us responsible, as well as the idea that, if this thing were replaced by the actions of lasers or parasites, that would be enough to remove our responsibility (and cause those observing us to revoke a responsibility attribution if they knew about the deviancy in question).

Conclusion

Our goal in this chapter has been to provide an introduction to our executive theory of responsibility, by laying out its basic tenets in a clear way to allow comparisons to the existing approaches. In the chapters that follow, we will continue to make these comparisons, arguing that our approach outperforms the competition by providing treatments of specific cases that are more sensible, more empirically testable, and more consistent with our existing folk psychology, reactive emotions, and legal apparatus.

We agree with Fischer and Ravizza that control of the right sort is crucial to responsibility. We have postulated that control over cognition and action, as exerted by the executive functions, is the specific type of control that humans operate with. Neither the reasons-responsive theorists nor the real-self theorists have attempted to discern how exactly the brain achieves reasons-responsiveness, or how the brain generates action in keeping with a considered, or second-order, belief. One benefit of doing this is to inform institutional responsibility assessments, such as those made in the criminal law, by making clear what sorts of scientific evidence may be relevant to criminal responsibility. In the next chapter we will develop a conception of criminal responsibility based on executive function. Then, in chapters 5 through 7 we will tie our theory strongly to existing results from the cognitive sciences, primarily by contrasting it with an existing theory of responsibility that is also empirically based, Neil Levy's consciousness theory.

4 Criminal Responsibility

In this chapter, we will argue that the structure of the criminal law reveals that the folk concepts grounding attributions of criminal guilt refer to operations of the executive functions. We will claim that H.L.A. Hart's capacity responsibility can be understood as requiring that a defendant possess an MWS (minimal working set of executive functions), that legal excuses involve underdeveloped executive functions or executive dysfunction, and that gradations in mens rea requirements can be understood as requiring different levels of involvement of executive processes.

Why Focus on Criminal Responsibility?

In the subsequent chapters we plan to focus our analysis on criminal responsibility with the aim of providing a convincing account of canonical or paradigmatic cases of responsible action and excuse. Many other philosophers interested in responsibility employ a different tactic: David Shoemaker, for example, specifically argues from cases "on the margins"—those in which we are unsure about an agent's responsibility—to a pluralistic approach to moral responsibility, arguing that our ambivalence in such cases can be explained by different concepts of responsibility pulling against each other (Shoemaker 2015). John Doris's theory of responsibility is similarly focused on marginal cases. Indeed, he argues that the class of canonical cases is smaller than most think, and the class of marginal cases relatively large (2015). Work by responsibility theorists Nomy Arpaly (2003) and Angela Smith (2005, 2008) also focuses on understanding difficult, marginal cases where unconscious or non-deliberated attitudes and decisions can be linked to responsible action.

We agree that a theory of responsibility should provide an acceptable account of hard or marginal cases. However, we are also mindful of the legal axiom "Hard cases make bad law." That is, sometimes focusing on cases at the margins can obscure our understanding of the general rule that might cleanly capture a large proportion of the cases. Our goal in this book is to explore the folk-psychological mental states and processes that underpin responsible action and to uncover the brain processes behind these mental states and processes. By focusing on criminal responsibility—a type of responsibility for which there are high epistemic standards due to the serious consequences of criminal guilt—we aim to provide an account of the clearest and best-grounded cases of responsible action, which can reasonably be extended to the hard cases.

Once the scaffolding of that account is in place, we will examine whether our theory can help us better understand and classify certain hard cases of responsibility and categories of excuses relevant to the criminal law, with the goal of understanding which factors may excuse persons from responsibility. We will do so wary that folk responsibility assessments of such difficult cases—involving, for instance, psychopathy, juvenile status, schizophrenia, or sleepwalking (or one of its variants, REM behavior disorder)—might be flawed in a number of ways, and in some cases, less reliable. If responsibility attributions are grounded primarily in reactive attitudes, as many philosophers claim (Strawson 1962, Vargas 2013, Shoemaker 2015) and as we argued in chapter 3, then we might expect such human sentiments—such as moral anger—to be fairly reliable in the way they demarcate typical, robust practices of human agency, but less reliable in the way in they classify more marginal cases. As Shoemaker notes, in hard cases, reactive attitudes grounding responsibility (say, reactive attitudes assessing character or judgment) can pull in opposite directions; for example, one might feel moral anger regarding the act but not the actor's character (Shoemaker 2015). Just as there is general agreement—at least within the Western world—that both ice cream and French fries are tasty, but not that horseradish or eel is tasty, so too might folk reactions to central cases of responsible action converge in some cases but diverge in hard or marginal cases of responsibility. For example, as Shoemaker claims, cases where an actor is a psychopath seem to involve reactive attitudes and folk notions of responsibility pulling in different directions, both within persons and across persons. I may feel the psychopath is not responsible for being a bad person but is responsible for acting in a way harmful to others; you may feel he is responsible for both.

As indicated by prior chapters, this book offers a scientific understanding of the cognitive capacities necessary in many, and probably most, cases of moral responsibility. Cases of criminal responsibility overlap significantly with cases of moral responsibility. If Thomas breaks into his neighbor's house to steal things, he is morally responsible. He is also criminally responsible, assuming he is arrested and a prosecutor can prove Thomas committed a theft beyond a reasonable doubt. In Thomas's case one might see criminal responsibility as layered and possibly dependent upon Thomas's moral responsibility: In many cases criminal responsibility amounts to moral responsibility plus fulfillment of the criteria necessary to build and prove a legal case.

We acknowledge that moral and criminal responsibility do not always overlap. Failing to stop completely at a stop sign in the middle of night on a deserted road is an act for which a person is legally, but maybe not morally, responsible. Cheating on a spouse is an immoral act, but not an illegal one. We apply moral blame to persons for thoughts alone—we may blame a person for consistently engaging in lustful thoughts about someone other than his or her spouse—although the criminal law never holds a person responsible for thoughts alone. Even so, we agree with David Brink and Dana Nelkin that the two concepts of responsibility have "similar structure" (Brink and Nelkin 2013, 284) and think the capacities needed to understand and behave in accordance with both moral and legal rules rest upon the same brain structures and functions. Further, we will argue that criminal punishment rests primarily upon moral desert. Thus, the capacities for moral agency and legal agency intersect significantly.

Criminal Responsibility and Folk Notions of Responsible Agency

The criminal law hews strongly to folk-psychological conceptions of responsible agency and, as such, provides a vast and reliable repository of such conceptions. Here we will primarily refer to the criminal law of common-law countries such as the United Kingdom, the United States, and Australia. The criminal law of these countries contains cumulative folk wisdom, in a sort of crystallized form, regarding the psychological capacities and states an agent must possess to be held responsible for causing harm. These folk crystallizations have been tested in the face of evidentiary standards and burdens of proof designed to ensure that their application in criminal cases is reliable. Because persons deemed responsible suffer punishment—often consisting in a denial of liberty via incarceration—criminal verdicts have very high stakes. In the United States there are many safeguards aimed at preventing false verdicts, including (1) those guaranteed in the U.S. Constitution (including the rights not to incriminate oneself and to be represented by an attorney); (2) the axiom that a defendant is considered innocent until proven guilty; (3) the rules of evidence in criminal procedures (such as rules against hearsay evidence); and finally, (4) the "beyond a reasonable doubt" standard of proof. Civil cases are often determined by a "preponderance of the evidence" standard, but under the criminal law, a defendant must be found to be a responsible agent who caused criminal harm such that a normal observer with access to all the relevant (legally acquired) evidence would believe without any reasonable doubt the defendant was responsible. This high burden of proof and the other safeguards reflect the seriousness of our extended folk practices used to assess an offender's responsibility.

We contend that the conceptions of responsible agency found in the criminal law—derived from the folk and supported by reactive attitudes—are stable and robust indicators of the capacities required for responsible action. In the common-law tradition, historical treatment of responsibility by the courts coalesced into precedent, used along with criminal codes that guide the current handling of criminal cases. Criminal codes specify what law enforcement, prosecutors, judges, and juries should look for in determining (1) whether the accused is a person with the capacities for responsible agency (or whether his or her actions are excused), (2) whether this person caused the criminal harm alleged, and (3) whether this harm was justified in the eyes of the law. Looking for specific mental capacities helps ensure that the offender has no excuse available since many forms of excuse are categories of psychological incapacity. The second criterion is directed at determining whether the specific elements of the crime are met and discerning whether this particular offender committed a voluntary action that satisfies these elements. The third assumes mental capacity and that a crime has been committed by the offender but looks to see whether the offender's act might be justified, as is the case with self-defense, for instance.

A small handful of philosophers have argued that the assumptions behind the law's use of folk psychology may be radically false (Caruso 2016, Pereboom 2013, Levy 2014). However, we accept that the folk-psychological states attributed to defendants do indeed refer to real mental states and processes in the defendants' heads (see Sifferd 2006). We also accept that the overall project of attributing blame and punishment to criminal defendants is not premised on a false claim about human decisions or behavior; it does not depend, for instance, on defendants having a type of free will or authorship of their actions that does not exist, as claimed by Pereboom (2013) and Caruso (2016). As discussed in chapter 3, we are compatibilists about responsibility, meaning we think that despite the laws of physics and our increasing ability to understand the mechanistic, causally determined nature of the physical underpinnings of human actions and decisions, we are still responsible for such actions and decisions (Stace 1953, Ayer 1969, Vargas 2013, Murray and Nahmias 2014). This is not to say that the criminal justice system is correct in every assumption regarding human psychology or behavior or that the generation of criminal verdicts and punishments is always fair or just. Instead, to get our theory off the ground, we need claim only that the folk psychology the criminal law rests upon is basically valid—that folk concepts of mental states and processes refer to underlying brain states and processes, and the attributions of blame and punishment made in the criminal law do not necessarily require the capacity for contra-causal (or extra-causal) free will, or some nonexistent human capacity.

The Components of Criminal Responsibility

Below we will provide a brief overview of the structure of the legal conceptions of criminal responsibility. We also hope to illustrate why the criminal law and criminal verdicts provide insight into paradigmatic cases of responsible action and excuse, couched in folk terms. As indicated above, we think exploring criminal responsibility can help us gain insight into the organization and reach of the folk concepts used to attribute responsibility in such canonical cases of responsible action, especially folk-psychological concepts.

The United States' Model Penal Code (MPC) defines both the voluntary act (actus reus) and the mental state (mens rea) requirements that must be met for a defendant to be found guilty of a crime. According to the voluntary act requirement, "a person is not guilty of an offense unless his liability is based on conduct which includes a voluntary act or the omission to perform an act of which he is physically capable." The MPC then lists the following as failing to qualify as a voluntary act: (1) a reflex or convulsion; (2) a bodily movement during unconsciousness or sleep; (3) conduct during hypnosis or resulting from hypnotic suggestion; and (4) a bodily movement that otherwise is not a product of the effort or determination of the actor, either conscious or habitual. Persons may only be held responsible for an omission if the omission is explicitly deemed sufficient by the law defining the offense, or a duty to perform the omitted act is otherwise imposed by law.

The voluntary act requirement is not solely an external bodily criterion; that is, an external observer might not be able to discern whether it was met without knowing something about defendant's past actions or mental states. Indeed, in some cases the requirement seems to be looking for a certain type of connection between the defendant's intentional states (e.g., the defendant's beliefs or desires) and the criminal harm. This link is certainly missing in cases of reflexes or convulsions, such as an epileptic seizure, and in cases of movement that is not the product of the effort of the actor, such as the wind blowing a man into the street. Movement during sleepwalking or hypnosis is more complicated. Sleepwalkers and the hypnotized are capable of very complex actions that indicate that perceptions of the world are playing a role. For example, sleepwalkers rarely run into walls, and they can open doors, prepare meals and eat them, and even drive. Similarly, persons can perform fairly complex tasks under hypnosis. But neither sleepwalkers nor persons under hypnosis are acting in accordance with desires, beliefs, or intentions produced by a cognitive system such that the actions "belong" to the actor. Instead, we readily say that the sleepwalker or person hypnotized didn't perform their acts "on purpose" or voluntarily—hence, they fail to meet the voluntary act requirement. In chapter 7, we will discuss such cases in detail; in brief, we argue that what matters to cases of sleepwalking or hypnosis is that agents have no possibility of subjecting their actions to "top-down" or executive review. If agents in those states cannot review their actions in light of their plans, goals, values, memories, and so forth and inhibit behavior on that basis, then those actions—however complex—are not ones for which the agents are responsible.

The mental state requirement looks specifically at the nature of a defendant's mental states in relation to the criminal harm caused. The MPC sorts such mental states into four categories that exhibit the principle of correspondence, meaning that the different mental states assigned to defendants reveal a sliding scale of culpability.[1] Defendants are most culpable for acts committed for the purpose of causing criminal harm ("purposely"), and slightly less culpable when they did not act for the purpose of causing harm but knew harm was likely ("knowingly"). Defendants are somewhat culpable when they consciously disregarded a substantial risk to harm ("recklessly") and are minimally culpable when they should have known they were violating a legally required standard of care ("negligently").

A first-degree homicide might be committed in which the death was caused "purposely," for example, where the shot was fired for the purpose of killing the shooter's correctly identified target. However, one might also commit a first-degree homicide "knowingly," for example, where someone sets fire to a house for the purpose of getting the insurance money, knowing that a person is sleeping inside. A "reckless" homicide, sometimes called manslaughter, might include a case where the defendant consciously disregarded the risk that someone might be killed, for example, if he shot a gun straight up into the air above a crowd. Finally, a negligent homicide is an unintentional killing in which the defendant should be aware of the risk of death, where failure to understand that risk involves a gross deviation from the standard of care that a reasonable person would observe in the actor's situation. These differing degrees of culpability have heavy practical implications: in the United States, sentences for first-degree murder can run from 25 years in prison, to life in prison, to the death penalty; a sentence for second-degree homicide or manslaughter can run from 6 to 50 years.

At trial, judges and jurors hear evidence offered to prove both that the defendant committed a voluntary act that was a proximate cause of the criminal harm and that the defendant possessed the requisite mental states for the crime charged. A proximate cause is understood to be one which, in a natural and continuous sequence unbroken by any efficient intervening cause, produces some effect, and without which the result would not have occurred. Eyewitness testimony or electronic surveillance evidence may be offered to prove that it was indeed the defendant who committed the crime. Defendants rarely testify in their own criminal cases—for fear they may incriminate themselves on cross-examination or be made to look guilty—so most of the evidence offered is behavioral; rarely do defendants confess (if they do, the case is unlikely to go to trial), and defense witnesses cannot include what the defendant told others about the crime because of "hearsay" restrictions. The prosecution may present evidence that the defendant was seen intentionally pointing a gun at the victim, made plans to commit the crime, attempted to cover up the crime, or had motive to commit the crime. Motives include reasons the defendant would want the criminal harm to occur; for example, because the defendant would secure a financial or emotional advantage. The court is asked to use this information to determine what mental states the defendant possessed when the crime was committed.

In most cases a defendant found guilty of a crime is thought to possess *conscious* mental states regarding the criminal harm he or she caused. This would seem to be the case where a defendant acts for the purpose of causing a certain harmful result, or with the knowledge that the harm was likely to occur. However, in negligence cases criminal guilt may be assigned even though the defendant seems to have possessed no conscious mental states with regard to the criminal harm. The law assigns culpability to a defendant who violates an important standard of care even when the defendant lacked conscious awareness that the criminal harm would occur. For example, a theater owner who never consciously entertained the idea that his old electrical wiring would catch fire can still be found guilty of a crime if he fails to satisfy important safety codes and people die when an electrical short causes the building to burn down. The MPC negligence standard applies if the owner *should have known* of a substantial and unjustifiable risk.

This attribution of culpability to persons who do not consciously desire, or even consciously acknowledge the risk of, criminal harm is a very interesting aspect of the criminal law and a good example of criminal law practices' role as a source of data for, and even as a constraint upon, theories of responsibility. Negligence is an important legal category, and assigning responsibility to those who should have been aware of a substantial risk seems to be an important folk practice not just present in the law but in our everyday lives. In chapter 5 we will discuss in detail the role that consciousness plays in responsibility, and in chapter 7 we will try to provide support for attribution of responsibility where agents are not consciously aware of the moral implications of their action. In doing so, we will distinguish our theory of responsibility from another prominent theory of responsibility informed by the cognitive sciences, that of Neil Levy (2014), who argues that consciousness of the moral implications of one's action is necessary for responsibility.

There is a final, somewhat neglected aspect of criminal verdicts: The prosecution in a criminal case must be careful to prove the right sort of causal link between a defendant's culpable mental states, his or her bodily movements, and the criminal harm for which he or she has been arrested. Certain very odd cases highlight the need to establish, and not simply assume, this link. Imagine a case where a defendant, Andrea, formed an intention to kill her neighbor, Jessica. Andrea's behavior causes Jessica's death; however, Andrea had no intention of killing Jessica at the time Jessica actually died. Andrea intended to kill Jessica in the late evening by poisoning the bottled water left on her porch. Instead, Andrea actually causes Jessica's death in the midafternoon when she sets her own woodshed on fire (unbeknown to Andrea, Jessica is hiding inside). This killing occurs a full five hours before she intends to poison Jessica's water—Andrea did not set fire to her woodshed with the intention of killing Jessica. In this case Andreas's action that resulted in the death of Jessica is not causally related to Andrea's intention to kill her. Thus, Andrea cannot be held responsible for Jessica's murder (although she may have been negligent in the way she set fire to her shed).

The next section concerns legal theories of responsibility. Specifically, we will examine H.L.A. Hart's notions of both capacity responsibility and legal liability responsibility. Later in the chapter we will attempt to understand criminal responsibility—specifically, capacity responsibility—in terms of control-based or "reasons-responsiveness" theories of responsibility (Fischer and Ravizza 1998, Vargas 2013) although, as discussed in chapter 3, we feel our theory may also be compatible with certain versions of the deep-self view. We believe it is no surprise that the folk concepts underpinning the criminal law identify the cognitive functions that give rise to sensitivity to important reasons for action (such as the law) and volitional control, the same functions required for responsibility under a "reasons" account such as that advocated by Vargas (2013).

Capacity Responsibility

H.L.A. Hart, perhaps the most influential legal scholar of the twentieth century, argued that legal responsibility refers to two different types of responsibility: the first type, which Hart calls "capacity-responsibility," consists of the general qualifying conditions for criminal responsibility (Hart 1968). The second, which Hart calls "legal liability-responsibility," consists of the specific mental state attributions the criminal law must assign to a defendant if he or she is to be found guilty of a particular criminal act. In addition to requiring specific mental states, legal liability responsibility also requires "causal or other connexions between a person and harm, or the presence of some relationship … between different persons" (Hart 1968, 221).

We mentioned the components of legal liability responsibility above when we discussed the voluntary act and mental state requirements for a guilty verdict. But we have not yet discussed the preconditions for attributing criminal responsibility to an agent, which Hart deems capacity responsibility. Hart viewed capacity responsibility as necessary for legal liability responsibility, arguing that because capacity responsibility concerns a general ability to understand and conform one's behavior to rules, it is a foundational requirement for the efficacy of law. That is, the institution of law depends on persons being capable of understanding the rule of law and making decisions which abide by legal rules. If no such persons existed, or not enough such persons existed, then the institution of law would fail. Similarly, if a person or class of persons cannot perceive the law as a reason to act and conform their behavior to it, the law fails as applied to that person or class because the law cannot influence their behavior (for instance, if they are seriously mentally ill); they fall outside of its reach. In sum, it is not fair or just to claim that a defendant has satisfied the mental states requirement for guilt unless that defendant has the capacity to recognize and behave in accordance with legal and moral rules.

Hart's discussion of capacity responsibility is an important first step toward understanding the cognitive capacities that underpin responsible action under the law. Although Hart was writing before the advent of contemporary neuroscience, the capacities he deems necessary for criminal responsibility are strikingly similar to the executive processes described in chapter 2. Hart claims that such capacities include "understanding, reasoning, and control of conduct: the ability to understand what conduct legal and moral rules require, to deliberate and reach decisions concerning these requirements; and to conform to decisions when made" (Hart 1968, 227). These capacities may be "diminished" or "impaired" as well as wholly lacking, "...and persons may be said to be 'suffering from diminished responsibility' much as a wounded man may be said to be suffering from a diminished capacity to control the movements of his limbs" (Hart 1968, 228). One legal responsibility scholar, Antony Duff, says Hart's capacity responsibility "...specifies a minimal condition of liability, which is satisfied alike by a willful murderer and by one who negligently causes death. That minimal condition could, however, be taken to require the possibility of effective choice: a person has the capacity to obey the law, we can say, only if she would obey the law if she chose to do so, and has a fair opportunity to obey only if she has a real choice of whether to obey or not" (Duff 1993, 347).

Hart's analysis describes two different types of human capacities he deemed necessary to legal agency: higher-level cognitive capacities and volitional control. Very generally, high-level cognitive or reasoning capacities allow persons to understand the demands of the law and the relationship between their behavior and the law, and volitional capacities allow persons to conform their behavior to the law's demands. If a defendant lacks capacity, that defendant may claim a criminal excuse, such as juvenile status, legal insanity, or diminished mental capacity. In some cases, a lack of capacity responsibility must be proven via evidence offered to the court, and in other cases, it is assumed given some reliable indicator. For example, the MPC states that a person shall not be criminally tried or convicted if at the time of the conduct he was less than 16 years of age. This is because the writers of the MPC assumed persons under 16 lack fully developed mental capacities for reasoning and volition and thus gave them a categorical excuse, guaranteeing that juvenile offenders were deemed less culpable and less deserving of punishment. For defendants under 16, the MPC gives exclusive jurisdiction to a dedicated juvenile court, designed to be less punitive and more focused upon rehabilitation.

Hart's notion of capacity responsibility has been hugely influential, and its echoes are felt in many contemporary theories. For example, David Brink and Dana Nelkin (2013) claim that the "architecture" of responsibility—whether legal or moral—requires cognitive and volitional capacities. On a broadly retributive view, blame and punishment are thought to be fitting responses to culpable wrongdoing. Such blame and punishment, however, is only fitting if an offender had "fair opportunity to avoid wrongdoing" (Brink and Nelkin 2013, 284). Certain cognitive and volitional capacities provide this fair opportunity because they enable "normative competence": They allow an offender to understand moral and legal rules and exert control over his or her behavior to avoid breaking such rules (Brink and Nelkin 2013). To see the importance of normative competence, consider the distinction between a justification and an excuse: Whereas a justification denies that a criminal defendant has committed a wrongful act, an excuse admits wrongdoing but offers reasons why the defendant is not culpable for such wrongdoing. Excuses require substantial impairment of either cognitive or volitional capacities, and it must be the case that the defendant is not substantially responsible for his or her own incompetence.[2] Psychopaths, for example, do not lack cognitive capacities according to Brink, although he claims it is possible (although unlikely) that some may lack the impulse control required to conform their behavior to legal and moral rules. However, mental illness may excuse a defendant if as a result of the illness the defendant is unable to understand the dictates of the law in relation to his or her behavior, or if he or she lacks the capacity to conform his or her behavior to those dictates.[3]

Discussions of cognitive competence have tended to focus on the capacity to know whether an action or its result violates a moral or legal norm; Brink and Nelkin argue, however, that because causing criminal harm is justified under certain "choice of evil" circumstances, recognition of a legal norm requires that a person be able to grade offenses as greater and lesser. For example, Felicia may be justified in taking another's suitcase with the intention to permanently deprive its owner of its contents if she knows those contents include a bomb that will blow up an airport. However, she is probably not justified if she merely thinks the bag has fruit in it that may carry an unwanted foreign pest (assuming she is not working for the U.S. Department of Agriculture). Recognition of wrongdoing, therefore, involves understanding the nature and degree of seriousness of one's act in relation to the law, as well as in relation to the circumstances surrounding the act (Brink and Nelkin 2013).

To sum up, the general mental capacities necessary to culpable action under the criminal law fall into two general categories; higher-level reasoning capacities and volitional capacity. Only if a defendant has these capacities will she be fully responsible, even if she has the specific mental states necessary for criminal guilt. The voluntary act requirement demands that a culpable act be performed in relation to desires and beliefs produced by a cognitive system such that the action is consistent with the actor's beliefs and desires (and thus "belongs" to the actor). The mental state requirement dictates that we look for specific mental states with regard to the criminal harm caused by the actor. If a defendant desired the criminal harm, or knew it would result from his or her actions, the defendant is highly culpable for that harm. However, if a defendant did not consciously desire the harm, but knew his or her behavior involved a substantial risk of criminal harm, or should have known his or her behavior entailed a substantial risk of criminal harm where the defendant also violates an important standard of care, the defendant is also criminally culpable, but less so than if the harm was desired.

An Account of Criminal Responsibility Based on Reasons-Responsiveness

Fischer and Ravizza's reasons-responsive model of responsibility, discussed in chapter 3, seems to cohere nicely with the capacities emphasized by Hart, Duff, and Brink and Nelkin as necessary to responsibility under the law. Hart and Brink and Nelkin specifically highlight (1) higher-level reasoning capacities, including the capacity to understand moral and legal rules, and (2) volitional capacities as necessary to legal agency. These capacities overlap significantly with the capacities underpinning reasons-responsiveness. Remember that Fischer and Ravizza claim that a person is morally responsible for an action only when the mechanism issuing that action is *moderately* reasons-responsive. Moderate reasons-responsiveness requires (1) that an agent must be regularly receptive to a range of reasons, including moral reasons, such that the agent manifests an intelligible pattern of responsiveness over time; (2) that an agent held responsible for a particular act would react to at least one sufficient reason to do otherwise than he or she did in some possible scenario; and (3) that the agent have come to own his or her reasons-responsive mechanism by means of taking responsibility for it (Fischer and Ravizza 1998, 207–39). A reasons-responsive actor is receptive to a range of reasons for acting—including those connected to moral and legal rules, which are arguably some of the most important reasons for action—and has the capacity to exhibit this sensitivity both over time as a pattern of behavior and with regard to a particular act.

Capacity responsibility, as envisioned by Hart, appears to require moderate reasons-responsiveness. Again, Hart claims capacity responsibility involves "understanding, reasoning … the ability to understand what conduct legal and moral rules require, to deliberate and reach decisions concerning these requirements" as well as act upon them (Hart 1968, 227). Fischer and Ravizza claim that a responsible agent must exhibit a *pattern* of reasons-responsiveness, meaning that an agent who is moderately reasons-responsive has an appropriately sensitive mechanism that responds to reasons that hang together rationally as a class and fit a coherent pattern. This indicates that an agent is capable of both recognizing reasons—including legal and moral reasons—and acting in accordance with those reasons over time, which in turns means the agent possesses diachronic and often synchronic reasons-recognition, as well as volitional control.

Hart's claim that capacity responsibility specifies a minimal condition for liability also indicates that legal agency rests upon reasons-responsiveness. As noted above, the law cannot influence the behavior of persons who cannot grasp the law as a reason for action and conform their behavior accordingly—meaning, in Hart's view, the law cannot meaningfully apply to such a person. Reasons-recognition and volitional control over action are thus vitally important to the function of law. As we shall see below in our discussion of negligence, both synchronic and diachronic reasons-responsiveness underpin criminal liability responsibility. While such capacities exercised in a synchronic way may ground an act committed for the purpose, or with knowledge, of causing criminal harm, criminal negligence rests largely upon an agent's diachronic capacity to be law-abiding. That is, criminal responsibility for unforeseen results of an action issued by a reasons-responsive mechanism is grounded in an agent's diachronic capacities to be law-abiding. We will say more on this below and in our discussion of Levy's theory of responsibility in chapter 7.

It thus seems clear that the capacities cited by Fischer and Ravizza as necessary to reasons-responsiveness overlap significantly with those required by Hart for criminal responsibility. A deep-self account of responsibility, however, fits less well with the capacities legal scholars claim are necessary for responsibility under the criminal law. The difference between the two views is whether it is the ability to control our actions, as claimed by a reasons-responsiveness view, or the relationship between the self and action, as claimed by a deep-self view, that matters most to responsibility. The discussions of the capacities grounding criminal responsibility by Hart (1968) and Brink and Nelkin (2013) make clear that volitional control is necessary, which coheres nicely with Fischer and Ravizza's notion that a person must have diachronic and synchronic capacities to act in accordance with reasons to be responsible. Whether an act that results in serious criminal harm expresses the actor's real self, however, is generally irrelevant to capacity responsibility. The law doesn't much care if a murder is wildly out of character for the defendant, so long as he had the capacity to understand moral and legal rules and conform his actions to such rules at the time the murder is committed. Similarly, in a negligence case, whether it was out of character for a defendant to forget to have his truck's faulty brakes fixed is not relevant to his culpability—unless there is additional evidence that the failure was a symptom of his lack of capacity for reasons-responsiveness at the time of lapse.

The Brain Processes and Functions Necessary to Capacity Responsibility

Our theory offers an account of criminal responsibility and responsible agency that is capacitarian and compatibilist. It is broadly a reasons-responsiveness account, but one that is explicitly brain-based, using cognitive science (especially neuroscience) to inform our model of the capacities necessary for responsible action. Specifically, we attempt to understand Fischer and Ravizza's reasons-responsive mechanism, and the capacities Hart claims are necessary to legal agency, in terms of brain function.

The primary thesis of this book is that the folk and legal concepts underpinning the structure of criminal offenses and verdicts discussed above implicitly refer to a particular set of cognitive functions that reside primarily in the prefrontal lobes of the brain and are understood by cognitive science as executive functions.[4] Examination of folk judgments about typical or normal cases of legal responsibility, as well as analysis of cases at the margins of excuse, indicate to us that legal responsibility is contingent upon a person's having the capacity for some baseline level of executive function—what in chapter 3 we called a "minimal working set" of executive functions. We will argue that the law implicitly assumes that citizens possess some baseline executive capacity, and that cases of legal responsibility are cases in which an offender either exercised executive functions with regard to the criminal act or should have done so (unless some justification applies). We will further argue that cases of legal excuse tend to be cases where an offender had severely compromised executive functions, either chronically, as in the case of brain damage specifically affecting the executive processes, or acutely, as in the case of extreme intoxication or crimes of passion, both of which involve executive malfunction. Finally, in some cases where a defendant's executive processes did not interact with information regarding the criminal harm caused, such as in cases where a lapse causes the harm, the defendant may still be responsible under the principle of negligence based upon the defendant's possession of an MWS within a broader time frame running up to the lapse, and his or her failure to meet a legal standard of care that reasonably should have been met. Diachronic capacity responsibility, via possession of an MWS, grounds the law's demand that individuals organize their minds and environments to meet certain legal standards of care, explaining and justifying why the law may attribute culpability in lapse cases.

Consider once again the cognitive capacities identified as necessary to capacity responsibility by Hart: understanding, reasoning, and control of conduct; the ability to understand what conduct legal and moral rules require, and to deliberate and reach decisions concerning these requirements; and the ability to conform to decisions when made (Hart 1968, 227). These mental capacities, which we have argued include higher-level reasoning capacities and volitional control, implicitly identify the suite of executive functions described in chapter 2. Executive functions—including planning, attentional control, inhibition, and task switching—are what allow us to navigate a complex world by acting in accordance with reasons, plans, and values. Together, they constitute the core of the reasons-responsive mechanism that grounds moral responsibility (Fischer and Ravizza 1998, Vargas 2013). As argued in chapter 3, executive functions allow agents to recognize moral and legal norms as reasons for action (or inaction) and to exercise the self-control to abide by such norms and laws.

Thus, a primary claim of this book is that actions for which we are responsible either involve executive processing or should have involved executive processing. Here again is our basic schema, now tailored to describe criminal responsibility:

Jo is responsible for a criminal illegal act, omission, or consequence only if

1. Jo has a minimal working set of executive functions (MWS), and
2. Jo performed the act or omission, and/or caused the consequence, or failed to act to prevent it, and
3. Jo's executive processes either played an appropriate role in bringing about the action, omission, or consequence or should have played an appropriate role in preventing it.

The first clause requires that Jo have capacity responsibility by way of possessing an MWS of executive functions. The standard of an MWS, developed in chapter 3, provides a workably definitive boundary of legal capacity. Although each person has his or her own executive suite of cognitive capacities with different strengths and weaknesses, the law must set a bright line regarding what constitutes enough capacity to ground responsibility. Our notion of an MWS accommodates interpersonal differences and acknowledges that intrapersonal weaknesses in certain executive functions may be compensated for by other executive functions. As we mentioned in chapter 2, current behavioral and neuroscientific tests are, considered individually, too coarse-grained (or too narrow in scope) to establish that a person has an MWS. Still, standards could be developed such that a collection of tests might reliably establish whether a defendant possesses an MWS and thus has capacity responsibility. This has very serious implications for the criminal law. Because capacity responsibility is the first hurdle to legal liability responsibility, a reliable empirical means for establishing capacity (and incapacity) should result in more accurate, and thus just, criminal verdicts and sentences.

The second clause stipulates that Jo's act or omission is voluntary in the legal sense, which means it is loosely related to Jo's agency such that it is related to Jo's mental states, and that the action or omission is an identifiable proximate cause of the harm. The third clause describes the nature of the connection between Jo's criminal responsibility and his set of executive processes. In many cases, a defendant's MWS will have played a direct role in causing the behavior via planning, attending, evaluation of consequences, error correction, or the like. Where a crime requires that a defendant acted *purposely* or *knowingly*, for example, executive functions must be a direct cause of the criminal action. In cases where a crime is committed *recklessly*, a defendant must have consciously considered the risk of harm and then failed to act to diminish the risk. In some cases, however, despite the possession of an MWS, a defendant may fail to stop the act or omission from occurring when he or she could have done so (e.g., the defendant's MWS may fail to fully engage and thus fail to inhibit a hasty reaction or his or her MWS may fail to plan a complex action). In these cases, the defendant's MWS should have engaged at the time the defendant committed the criminal act, or sometime before, but failed to do so. In other cases the failure of a defendant's MWS to engage may be explained by the cultivation of habits resulting in the commission of a crime, or a defendant may fail to cultivate habits important to following the law. In such cases, capturing the agent's responsibility may require expanding the time frame during which the agent had an opportunity to use his or her MWS to be law-abiding. Because cognitive and volitional control can be spread over longer time spans—and are not limited to the short window just before an action—we must make room for diachronic operations of the MWS in assessing criminal responsibility, just as we did regarding moral responsibility in chapter 3.

In cases of excuse we claim that either a defendant fails to possess an MWS, or the synchronic or diachronic operations of the defendant's MWS are compromised, where the executive failures are not the fault of the defendant. For example, very young children do not have capacity responsibility because they do not yet possess an MWS. (We will discuss the emerging capacity responsibility of juveniles in depth in chapter 8.) In cases where a defendant is legally insane, it may be either that a mental illness has caused the defendant not to possess an MWS or that the illness is such that it is not possible for the defendant's MWS to be causally related to the criminal act in the typical way. In cases of mistake of fact, a defendant's MWS was also not related to the criminal act in the normal way, because of the defendant's (reasonable, nonculpable) mistake. In general, Jo needs to have a roughly accurate mental representation of certain important circumstances surrounding his action for him to have a fair opportunity for her MWS to produce a law-abiding action. Recall the mistake case discussed in chapter 3: If Jo is a hunter who shoots an animal activist wearing a bear costume in the woods, he is probably not culpable at all, compared to a scenario where he knowingly murders the activist under the cover of a hunting expedition. This is because in the first case Jo's executive activity, while obviously involved in the decision to shoot "the bear," was not causally interacting with a mental representation of a human activist. Jo's mistake was reasonable, so he did not have a fair opportunity to be law-abiding.

We will discuss the way in which our theory handles specific cases of excuse in subsequent chapters. If our theory is correct, evidence of executive capacity or incapacity may be particularly relevant to criminal trials. Over the past 15 years attorneys have increasingly introduced evidence of brain function, couched in terms of neuroscience or cognitive science, as relevant to criminal law verdicts (Farahany 2015). Courts must make the difficult determination whether such evidence bears on the folk concepts used to generate verdicts, and legal scholars have raised high-profile (and understandable) concerns about whether brain-based evidence is being properly understood and responsibly deployed (Morse 2006a). We believe our theory can be helpful, especially in determining what neuroscientific evidence may not be useful (e.g., evidence of brain dysfunction, such as perceptual problems, that can be corrected for via normal executive functions, and certain diagnoses that do not involve executive dysfunction). Even where our theory indicates neuroscientific evidence is relevant—for example, because it is evidence of diminished executive capacity—such evidence will need to be "translated" back into the folk concepts that generate criminal verdicts. Here again we think our theory can help, by acting as a bridge between information about a defendant's brain and the folk psychological concepts utilized to determine if he or she is criminally responsible. For example, in chapter 9 we will argue that neuropsychological evidence of serious executive dysfunction in persons with schizophrenia bears directly on the MPC's test for legal insanity, a test couched in folk terms.

Conclusion

As we acknowledged in chapter 2, scientific understanding of executive functions is still maturing, but already an impressive amount of work has been done in isolating them, testing them, and probing their neural bases. We believe a clear understanding of the way in which these functions give rise to responsible agency can help illuminate the folk understanding of responsible action. Further, once we understand that responsible agency rests upon executive functions, a scientific understanding of executive functions might allow us to recalibrate our perspective on some difficult cases in the criminal law, where complex and counterintuitive phenomena outstrip our folk-psychological language. Folk concepts in the criminal law relating to the offender's psychology are fairly coarsely grained by necessity: Folk hypotheses regarding mental capacities are categorical and driven largely by observation of outward behavior (with help from our mind-reading capacities). Understanding these folk concepts in terms of executive function may allow us to offer solutions in cases where the folk concepts have been too coarse to treat like cases alike (or different cases differently). We think agents such as those discussed in the case studies that begin chapter 1—juveniles, schizophrenics, and psychopaths—are the sorts of agents the folk have difficulty categorizing with regard to their responsibility. Now that we have a clear understanding of the cognitive grounds of responsibility in canonical or "core" cases, we can begin to critically address these hard or marginal cases of responsibility; in chapter 8 we will discuss juvenile responsibility, and in chapter 9 the topic of legal insanity and how mental illness relates to legal categories of excuse. Further, in chapter 7 we will discuss a case of criminal negligence and compare our treatment of a case with a close competitor, Neil Levy. We believe that our approach can inform the folk concepts of responsibility relevant in these cases and thus clear up some of the confusion surrounding these types of offenders.

But first, we will distinguish our approach from other scientifically based accounts of responsible action—specifically, those that claim that what matters most to responsibility is consciousness, not executive processes. This is the topic of the next two chapters.

5 Consciousness versus the Executive Processes

In this chapter, our goal is to distinguish the roles of executive control and consciousness in allowing us to be responsible. Our hypothesis is that when a competent person is responsible for doing something immoral or criminal, two conditions hold: First, that person played a crucial role in the causal chain(s) that resulted in the crime or the harm—that person was the one who pulled the trigger, grabbed the cash, or ran the red light and caused the accident. Second, the presence (or absence) of certain types of executive activity plays an equally crucial role in making a person responsible.[1] If the person intended the crime, if the person planned it, or if the person failed to stop the action from taking place when he should have, this mental condition for responsibility can be met. We made this claim specific and empirically testable by arguing in chapter 3 that the cognitive control network and its various executive functions are a vital part of the mental component of what makes people responsible.

We are mind–body materialists, or more specifically physicalists, and our approach understands these mental components required for responsibility as composed of brain states, events, and processes.[2] The statements constituting the more scientific portions of our theory are in the language of neuroscience. We are thus assigning a specific ontology or metaphysical status to mental states, the one described by the language of neuroscience, which employs concepts of entities such as *neuron, axon, white* and *gray matter, cortical areas,* and concepts of processes such as *activating, modulating,* and *inhibiting*.

Levy on the Role of Consciousness in Responsibility

Philosopher Neil Levy's (2014) book *Consciousness and Moral Responsibility* is an important landmark in the inquiry into human responsibility, as the first comprehensive theory explicitly based on findings in contemporary cognitive science. Levy sketches a theory of responsibility according to which a person is responsible for actions only when he or she is conscious of the moral implications of those actions. We will focus on Levy's theory in this chapter, as well as chapters 6 and 7, for a couple of reasons. First, Levy draws connections between his theory and scientific findings at several points. This allows his theory's interpretation of these results to be tested against our theory's interpretation in specific ways. Second, the theory is close enough to ours to make all sorts of informative comparisons and contrasts between the two possible. Levy's theory uses much of the same data from cognitive science that we rely on yet reaches a different conclusion. Levy bases his account on a theory of consciousness which is accepted, in different versions, by several of the most prominent scientific and philosophical theorists. Our account is based on an opposing theory of consciousness, also backed by prominent thinkers and supported, we will argue, by powerful new research paradigms and evidence.

The connection to brain science that both we and Levy make is important for several reasons. The standard approach in philosophy toward consciousness is to treat it as a sort of given that we all are familiar with in virtue of having a consciousness of our own or in virtue of having mastered the concept of consciousness. We think making the connection from philosophy to science constitutes a significant advance over this approach, in that it brings a vast range of new data into play that can be used to test and improve, or even conclusively reject, theories. Constructing an account like this more than doubles one's workload, though, since both the philosophical and scientific research literature must be surveyed, each of which are now vast. More than doubles, we say, because properly interpreting the scientific work involves entering fields where one is a novice and requires that one undergo a lengthy apprenticeship. In addition to the connection to science, Levy also shares our desire to create a theory of responsibility that coheres well with our ordinary ways of thinking and speaking about the mind, sometimes called folk psychology.[3]

The scientific theory of consciousness that Levy employs is accepted by several important research groups in neuroscience, psychology, and portions of the philosophy of mind. It is called the *global workspace theory*, or *global neuronal workspace theory*, when phrased in the language of cognitive neuroscience. According to this theory, consciousness functions as a broadcasting mechanism that relays its current contents to several brain subsystems, including the brain's memory systems, emotional systems, and the executive processes. We also accept aspects of this theory, but we will make several corrections to how the theory is formulated, and to how its creators have interpreted the experimental data. Specifically, we will argue below that existing misformulations or misstatements of the theory played a role in causing Levy to overvalue the role of consciousness in responsibility.

The global nature of the conscious workspace is what allows it to play its vital role in responsibility, according to Levy: "The integration of information that consciousness provides allows for the flexible, reasons-responsive, online adjustment of behaviors. Without such integration, behaviors are stimulus driven rather than intelligent responses to situations, and their repertoire of responsiveness to further information is extremely limited" (Levy 2014, 39). The connection to responsibility now follows, according to Levy: "[T]his entails that we possess responsibility-level control only over actions that we perform consciously, and that control over their moral significance requires consciousness of that moral significance."[4]

We disagree with Levy about the role of consciousness in making us responsible. Our claim is that the crucial thing is something else. The executive processes are controlling events in the workspace, performing various operations on conscious states; thus, we think they are the crucial factor. This disagreement is complicated, however, by the fact that we do not agree with Levy on the subject of what consciousness itself is. The two competing theory classes on consciousness point to different brain state types as being conscious states and employ different concepts of what consciousness itself is. We will argue that Levy and the others using the type of theory he favors are using "consciousness" in an overly broad sense that includes the executive processes themselves. Levy's failure to notice this causes him to assign many functions to consciousness that are actually executive functions. We plan to argue, further, that the two uses of "consciousness" employed by the two theories do not correspond to two different types of consciousness, and that there is only one type of consciousness. At several points below, we will make clear that this is more than a "mere semantic" dispute about what concept of consciousness to employ. Behind the dispute are two quite different accounts of both consciousness and the larger mental mechanisms needed to make us responsible beings.

Dueling Concepts of Consciousness

Philosopher Ned Block's distinction between two concepts of consciousness, which he called "access consciousness" and "phenomenal consciousness," has been widely used and figures importantly in Levy's account. So we need to describe it, although we will issue some caveats about how the distinction is made, and Levy has some of his own. According to Block, "[A] mental state is access conscious if you can think about it, report its content, or use it to guide action" (Young and Block 1996, 156).[5] Levy, helpfully, and we think correctly, relates access consciousness more closely to our issues here: "[I]nformation is access conscious when it is available to a broad set of systems, including systems involved in reasoning; when information is access conscious, it is also available for verbal report because it is available to the majority of systems driving behavior" (Levy 2014, 35). According to Levy, the concept of consciousness he is using, which he calls "personal availability," is coextensive with the concept of access consciousness.[6]

In contrast, phenomenal consciousness, says Block, is simply "the experience of seeing, hearing, feeling pain, etc." (Young and Block 1996, 156). We would correct this to note that, since experience assumes a subject of experience, phenomenal consciousness is not minimal enough to capture the idea of consciousness alone, rather than consciousness together with other processes, which we assume is our task here. If a subject of experience is included, phenomenal consciousness becomes too close to access consciousness, and too close to the notion of personal availability, because we have a subject with access to a conscious state. We prefer to employ a more minimal notion of consciousness, which we have called "bare consciousness," that refers only to the conscious state itself, with no implications about the presence of a subject, or the presence of functional connections to anything else in the brain for that matter, via broadcasting or any other causal connection.[7] However, we will ignore this issue and speak of phenomenal consciousness since it is the best-known term for referring to what we believe is the core phenomenon behind our concept of consciousness, a biological process that generates what we experience.

Levy is clear that he is not employing phenomenal consciousness in his theory, but rather, again, something quite close to access consciousness, which enables all of the functional connections between the distributed receiving mechanisms and the workspace. "The contents that constitute our identity are broadly distributed in the mind," and "consciousness enables the distributed mechanisms that constitute agents to play a coordinated and coherent role in that agency" (Levy 2014, ix). Levy explicitly de-emphasizes any important role for phenomenal consciousness in his theory.[8] "Consciousness is necessary for direct moral responsibility," claims Levy, "not because of what it *is*, but because of what it *does*" (ix).[9] This fits with the idea that phenomenal consciousness designates conscious *states* while access consciousness designates a type of functional role that consciousness stands in to other brain processes. Consciousness is more of a function than a state, as Levy sees it.

As we noted, Levy argues that something he calls personal availability is required for responsibility. Here Levy relates this concept to consciousness: "When an agent has such information personally available to her she is conscious of it," or aware of it (Levy 2014, 29, 34).[10] When information is personally available to a person, the person can "easily and effortlessly retrieve it for use in reasoning" and it is "online," in that it is currently guiding the agent's behavior (32–3). Once these criteria are met, the information that the agent has this relation to is "available to underwrite justified attributions of moral responsibility" (33n). People who have information personally available to them can also typically report it to others, according to Levy (29).[11]

Levy's theory is also aimed at capturing another important feature of responsibility: the idea that we are responsible for actions that express our attitudes. This is the feature of responsibility that the real-self theories we examined in chapter 3 are based on. Consciousness allows the current conscious state to take part in cognition along with the person's other attitudes, such as her beliefs and desires. "Only when a representation is conscious," says Levy, "can it interact with the full range of the agent's personal-level propositional attitudes" (Levy 2014, vii). Levy says, "Mental states are personal when they can be rightly predicated of the person herself ('Jenni believes that *p*,' for instance); and subpersonal when they can only be predicated of some lower-level components of the mind ('edge detection occurs in V1,' for instance)" (31n). The conscious workspace ensures that the current conscious state is globally broadcast, so that a complete survey can be made of any connections of interest between it and existing attitudes. This "entails that consciousness of the facts to which we respond is required for these facts to be assessed by and expressive of the agent herself, rather than merely a subset of her attitudes" (Levy 2014, vii).

It is not enough, however, merely to rouse the relevant attitudes and let them interact in the conscious workspace. We saw in chapter 2 that executive monitoring of memories, perceptions, and emotions (and the values they confer on people, things, and events) allows us to edit and rethink what we initially thought we remembered, perceived, felt, or even believed. Our attachments of moral significance to persons and events is a complex process involving emotional reactions, memories, current perceptions, and thought, including concerted bouts of thinking when inconsistencies need to be resolved. This entire process is carefully controlled and kept on point by the executive functions. Executive processes allow us to move beyond awareness of moral significance to *considered* awareness of moral significance. Our initial take on the moral significance of an action might be mistaken. For example, one might mistake social inappropriateness for immorality, and then realize one's error upon further consideration. Our considered moral significances express who we are much more truly than those that only occur to us spontaneously, in much the same way that, as Frankfurt pointed out, our higher-order desires reflect our real selves much more than our first-order desires (see chapter 3). Considered moral significances are those that have had the right sorts of engagement with the executive functions.

In what follows we will argue that Levy's emphasis on consciousness, as personal availability or access consciousness, is misplaced. Further, we will argue that the reason why Levy believes that consciousness is necessary to responsibility is largely that, as in the above example, he is attributing functions to consciousness that are actually accomplished by the executive processes. The executive processes allow for flexibility of response by adroitly manipulating perceptions, emotions, and memories. The executive processes are also the final arbiters of how workspace activity connects to the outside world via action. Our claim is that consciousness, understood properly, is itself a sort of display buffer, the contents of which are made available to several other brain systems, including the executive functions. A buffer such as this is analogous to the blackboard in a math classroom. As the class proceeds with a difficult proof, the board allows them to merge the minds of everyone in the room, by keeping everyone on the same page, as it were. A display buffer itself is not capable of coordinating the diverse group of brain processes that receive the broadcast at the cognitive level and combine their influences to produce actions, however. Executive functions are required. Once we clearly delineate what consciousness is and what executive function is, we will see that the concept of access consciousness, as Levy uses it, runs the two together.

Levy might respond that he has other reasons for giving consciousness an important role in responsibility. He says that "behavior driven by conscious states will be better integrated—more coherent and consistent—than behavior driven by nonconscious states" (Levy 2014, 64). While the process of consciousness itself, understood in the way we recommend, does perform operations that make conscious states more coherent and coordinated, it is the set of executive processes, functioning at a higher cognitive level, that control the mix of attitudes, memories, perceptions, and emotions that go into the production of coordinated, coherent, and ultimately responsible actions. The coherence and consistency that is provided at the level of consciousness itself is of a different variety. For example, visual processes fill in the visual blind spot in order to produce a gap-free visual field; they disambiguate ambiguous figures, such as the duck–rabbit; and they make colors appear constant and uniform (see, e.g., Walsh and Kulikowski 1998). But all of these "coherencing" operations occur before the conscious state is exposed to the executive processes and begins to play a role in cognition.[12] The brain processes associated with consciousness itself have a lot of work to do in producing this rich world representation in at least five modalities, getting it all filled in and consistent. However, that is different work from what the executive processes do, which occurs at the next level up in the cognitive hierarchy.

Global versus Local Theories of Consciousness

Before we can assess Levy's claims about the relevance of consciousness to responsibility, we need to get a fix on what consciousness is. Unfortunately, this issue is still very much up for debate. While there appears to be a majority view on what sorts of brain events constitute conscious states—the one Levy bases his theory on—the current science on consciousness is still unsettled. At the moment, there is a dispute between two large theory classes, without either one's attaining anything like a genuine consensus. According to the majority view, the global view, the presence of consciousness requires the operation of a large network of cortical areas, typically including posterior components in the parietal lobes, linked to prefrontal components, all of which are supported by numerous subcortical and brainstem systems. According to the opposing view, the local view, consciousness can occur in much smaller areas of cortex (which also require subcortical support, especially from certain thalamic nuclei). All that is needed for consciousness is the presence of a certain process in which large sets of neurons enter into a mutual oscillatory state, coupled with related parts of the thalamus, according to the localists.[13] In particular, the theorists in this minority camp agree that conscious states can occur in the posterior and inferior portions of the brain without active connections to associated prefrontal areas of the sort that the global theorists claim are required.

Our analysis will show that these two views are employing different concepts of consciousness. The global view is employing access consciousness whereas the local view is employing phenomenal consciousness (or bare consciousness). The proponents of the global view, we will argue, mistakenly lump the executive functions, or the cognitive control network, in with consciousness itself.[14] This suggests that access consciousness is not in fact a type of consciousness, but rather a broader type of mental state in which the executive processes are causally engaged with a (phenomenally) conscious state. The experimental work on this issue has recently reached a threshold level, so that this claim can be supported by a myriad of findings, which we will summarize below. We will argue that the experiments show that there are not two meanings of "consciousness," but rather just one, phenomenal consciousness, and that the uses of "consciousness" that lump it together with the executive processes are mistaken.

It is important to keep the intuitive functional notion of a workspace system separate from the specific remarks some of the workspace theorists have made about how consciousness relates to the workspace, as well as claims about what the neural underpinnings of the workspace system are. We accept the intuitive functional notion of a workspace while rejecting the idea that this commits us to a theory of access consciousness or to the neural systems pointed to by the global theorists as the grounds of consciousness. The intuitive notion of a workspace is plausible. We can observe the happenings in our own workspace, we are directly aware of what comes into consciousness and indirectly aware of executive activities being performed on them. We are aware of the participation of other processes that receive the global broadcast and respond. One clear example of this occurs when a conscious thought of something causes memories of that thing to come into consciousness. The idea of a workspace is neutral on the subject of access versus phenomenal consciousness; both types of theorists can make use of it although they point to different portions of the entire workspace system as being what "consciousness" refers to.

According to global workspace theory, the function of consciousness is to serve as a broadcasting mechanism, allowing the current conscious state to be accessed by a variety of different brain processes or subsystems. There are roughly analogous versions of the global theory in three different disciplines: psychology, philosophy, and neuroscience. The neuroscientific version is known as the global neuronal workspace theory or sometimes the frontoparietal theory of consciousness (Dehaene and Naccache 2001). The psychological version is known as global workspace theory (Baars 1993, 2002). One type of philosophical version is known as the higher-order thought theory of consciousness (Rosenthal 1986, 2005, Carruthers 2011).[15] According to the proponents of global workspace theory, the recipients of the broadcast include evaluative systems, attentional systems, perceptual systems, and long-term memory (Dehaene, Kerszberg, and Changeux 1998). In his recent book, Dehaene adds that "executive attention" receives inputs from the workspace, and "language, long-term memory, attention and intention systems are all part of this inner circle of intercommunicating devices that exchange conscious information" (Dehaene 2014, 167). The sensorimotor systems presumably also provide input about bodily movements, position, and so forth to the workspace. Some of the recipient processes also provide input back into consciousness. Our brains readily attach a sense of value to any contemplated scenario, for instance, and we can be consciously aware of that value.

According to localist theories of consciousness, on the other hand, a brain state can be a conscious brain state without possessing a functional connection to the prefrontal lobes and without becoming the object of a higher-order thought. Semir Zeki's has been the most widely read localist view among scientists (Zeki and Bartels 1999). The theory derives its name—microconsciousness—from its claim that there can be conscious states in small areas of cortex (perhaps with supporting subcortical areas). One good example of conscious states that routinely occur without active prefrontal connections, according to this approach, would be those parts of the visual cortex containing information from the unattended visual periphery. While you *can* attend to the periphery of your visual field, normally you do not. Yet those peripheral portions are still part of the larger conscious visual field that you experience. They are conscious states that you are not currently attending to or aware of. Granted, our ability to report the properties of stimuli in the unattended periphery is notoriously poor, but that does not imply that consciousness, or the conscious state, does not extend out to the periphery.[16] We are not seeing the world through a cardboard tube.

In a recent article reviewing the state of neuroscientific research on consciousness, Christof Koch and his colleagues (2016) assemble several strong arguments for the localist view, as well as a plausible explanation for where the globalists went wrong, all supported by recent research. The supporting studies they catalog involve new types of no-report paradigms, such as the use of measurement techniques that, unlike button pressing or verbally describing one's experience, will not activate the executive processes. These new techniques include measuring pupil dilation to see whether a stimulus has been detected, as well as using retrospective reports referring to conscious experience that took place several seconds before, such as asking subjects whether they were dreaming, upon awakening. A second approach involves showing that, as a cognitive task is made more difficult, the amount of executive activity increases yet varies independently of the activity required to support the conscious state itself (e.g., D'Esposito et al. 1999). These paradigms allow researchers to separate activity of brain areas containing executive processes from those that contain the conscious state itself. These latter exist in small "hot spots," in the vicinity of the junction of the temporal, parietal, and occipital lobes, according to Koch and his collaborators.

In one particularly telling type of comparison, the same study can be done with and without the requirement that the subjects make a report. When this is done, the report requirement produces activity in executive areas that was not present in the no-report condition. Binocular rivalry experiments have proven useful in this regard. Different stimuli are shown to the two eyes, forcing the brain's visual system to resolve the discrepancy as part of producing a single, unified visual field.[17] In one study using binocular rivalry, subjects showed widespread frontal-parietal activity (via fMRI) when asked to report what they perceived. However, when subjects experienced binocular rivalry but were not asked to make a report (a no-task paradigm), the neural activity in the frontal areas disappeared while activity that had been present earlier in occipital and parietal regions remained (Frässle et al. 2014).

Similar results were obtained in a masking paradigm in which blood flow to the crucial brain areas was monitored. Stimuli can be masked from conscious awareness despite being processed at lower levels of the visual system. Scientists can then study how altering the mask (or the stimulus) affects whether the subject can detect the stimulus. When subjects are asked to report whether they saw the stimulus or not, robust prefrontal activity appears. However, when the subjects are not asked to report, and physiological indicators such as eye movements or pupil dilation are used to determine whether the stimulus was seen, the prefrontal activity disappears (Pitts, Metzler, and Hillyard 2014).

Koch and his collaborators (2016) use the phrase "neural correlates of consciousness" (NCCs) to describe the brain activity required for conscious states, a phrase which we shy away from because of its dualistic connotations. Conscious states are not correlated with brain states; they *are* brain states. The authors make an important distinction, however, between "the content-specific neural correlates of consciousness" and the "full neural correlates of consciousness." They define the content-specific NCCs using an example of the neural correlates of "experiencing the specific content of a face." These neural correlates include the neuronal ensembles located in the inferior temporal fusiform face area that are active "whenever a person observes, imagines, or dreams a face" (Koch et al. 2016, 308). Then, they describe the full NCC as what all the specific NCCs have in common. The full NCC consists of "the neural processes supporting conscious experiences in their entirety, irrespective of their specific contents." Koch et al. (2016, 309) say the full NCC is located in a temporo-parietal-occipital hot zone.

Brain stimulation experiments, first performed in the 1950s, are also of use in demarcating executive activity from consciousness. Penfield (1975) discovered that it was possible to produce conscious states by stimulating the surface of the cortex with a tiny electrical current. Modern stimulation techniques are able to ensure that the electrical current does not spread beyond the target area, which supports the idea that the conscious states experienced by the patients exist solely in the stimulated area. In general, electrical stimulation to various posterior cortices produces conscious experiences, which begin with simple modality-specific sensations, such as flashes of color in early visual cortex, and get more complex as the stimulation moves in an anterior direction toward the central sulcus and the temporal poles. But this forward trend stops at the temporal poles: Stimulation of the prefrontal cortices does not produce conscious experience (see Selimbeyoglu and Parvizi 2010 for a historical review).

There are also clear examples from studies of neurological patients that support the localist approach. Patients with extensive frontal damage typically still show signs of consciousness. One patient gave clear signs of continuing to be conscious after near complete bilateral removal of the frontal lobes (Brickner 1952). The patient of Markowitsch and Kessler (2000) showed "massive impairment of executive function" due to severe degeneration of the prefrontal cortex but had "intact perceptual abilities and was obviously conscious," according to Koch and his coauthors (2016, 311). Severe autism is also characterized by underconnectivity between frontal cortex and posterior cortex, indicating that some people with autism may possess consciousness without robust prefrontal connections (Hirstein 2015).

There is also a new class of dream experiments that support the localist approach to consciousness if we make the assumption that dreams are a type of conscious state (more on this in chapter 7). Sleep-lab researchers awakened people at random intervals to ask them whether they were dreaming and, if so, about what. The study used a 20-second gap between measurement of any dream activity and the awakening. The researchers claimed that they were able to identify a good candidate for the full neural correlate of dreaming consciousness by using this task-free technique (Siclari et al. 2013). It is now well demonstrated that dreaming occurs not only during REM (rapid eye movement) sleep, but sometimes during non-REM sleep as well. Recently Siclari et al. have been able to show that activity in a posterior cortical hot zone predicts whether the sleeping subject will report dreaming, whether the subject is in REM sleep or non-REM sleep, suggesting, in the words of the authors, that activity in this hot zone "may constitute a core correlate of conscious experiences in sleep" (Siclari et al. 2017, 872).[18]

Further, there are numerous everyday experiences that, when brought into the laboratory, support localist theories of consciousness. When subjects are engaged in a task requiring perceptual concentration or are absorbed in watching a movie (ideally something like an exciting action movie that requires little thought), there is little prefrontal activity (Goldberg, Harel, and Malach 2006). Meditation may also provide another example of conscious states without prefrontal involvement. As in the case of perceptual absorption, meditation seems to shut down the cognitive control network while activity in the hot spots for consciousness remains robust. Deep meditation involves a significant diminution in prefrontal activity (Lou, Nowak, and Kjaer 2005) while consciousness continues.

Mistaking Executive Processes for Consciousness

What all this suggests is that the global neuronal workspace theorists have mistaken executive activity for part of the neuronal basis of consciousness, and that the core phenomenon of consciousness is something more basic, and smaller, than what they identified. Koch and his collaborators state it categorically: "The anatomical basis of the full NCC and content-specific NCC do not comprise the wide fronto-parietal network emphasized in past studies, but are primarily localized to a more restricted temporo-parietal-occipital hot zone with additional contributions from some anterior regions" (Koch et al. 2016, 315). The content-specific brain states involve hot-zone activity without the need for accompanying prefrontal-parietal activity, which Koch et al. describe as executive in nature: "Experiencing specific contents associated with activity in the posterior hot zone does not require the amplification of fronto-parietal network activity. Instead, the frontal cortex is involved in allocating attention and in task execution, monitoring and reporting" (315).[19] Koch et al.'s account of the error made by the globalist side is clear: The globalists have confused processes associated with the reporting, monitoring, and even inhibiting of conscious states with conscious states themselves. Using experimental paradigms that required report no doubt led to this mistake.[20]

Levy and Dehaene appear to be guilty of this error. They refer frequently to activity in prefrontal, parietal, and anterior cingulate areas as evidence for the existence of consciousness. This list of areas should sound familiar from chapter 2: Those are exactly the core areas that make up the cognitive control network. For example, Levy notes that Dehaene claims that the global workspace is realized by pyramidal cells in "prefrontal, parietal-temporal, and cingulate association cortices." The axons of these cells "form long-range thalamo-cortical loops allowing for the active maintenance of information and its availability to a broad range of the systems that drive behavior" (Levy 2014, 49–50). Later, Levy notes that frontal and parietal association cortex, and anterior cingulate cortex are "regions associated with consciousness" (73).

Thus, it appears that when the global theorists describe the anatomy of the global workspace, they are actually referring to the cognitive control network, possibly in addition to posterior areas that actually do contain the (phenomenally) conscious states, since these should also be showing up in their, for example, imaging data.[21] If this is correct, the globalists are, minimally, lumping the processes that operate on conscious states together with the states themselves, if they are not confusing cognitive control network activity with conscious states outright. Our view is that consciousness and the cognitive control network are not only separable, but that it is essential for theories of responsibility to separate the two, because it is not consciousness that is the crucial brain process for explicating responsibility, as Levy argues, but rather the set of executive processes, as realized in the cognitive control network. Consciousness and the executive functions are different processes, perform different functions, and exist in different brain areas.

What exactly is a *workspace*? There is a problematic ambiguity in the way the global theorists employ the term "workspace." The ambiguity occurs in the work of Dehaene and is taken up by Levy in his account of responsibility. Does the concept of a workspace include whoever, or whatever, does the work in the workspace? The proponents of the theory, including Levy, sometimes use the word "consciousness" to refer to both the workspace together with (some of) the processes that receive the broadcast and operate on its contents. At other times, however, they use "consciousness" to refer simply to the workspace itself, that is, whatever is "holding" the conscious contents that are being broadcast, analogous to the blackboard in that math classroom we mentioned earlier. The distinction is vital. Computer scientists, for example, take care to distinguish what they call "buffers" or "registers" that hold data to be operated on from the parts of the central processing unit that perform the operations. Mathematicians likewise distinguish between *operands*, such as numbers and variables, from *operators*, such as the plus sign, that operate on the variables. It seems reasonable, we suppose, to need to refer to the buffers together with the processes that operate on them as a single system—but not with the same word we use to refer to the buffer alone, to avoid causing vexing misunderstandings. We should thus take care to distinguish the two senses of "global workspace," the first referring to something like a buffer that contains the (phenomenally) conscious state (and only that), and the second referring to a larger system that includes processes that that causally interact with the buffer—such as the executive processes—together with the buffer. Here are definitions of the two: *workspace buffer* = cortical area(s) where conscious states exist; *workspace system* = the workspace buffer together with the set of processes that causally interact with it, including the executive processes.

Levy and Dehaene appear to use "workspace" in both of these senses, something that we believe kept them from seeing that they were sometimes referring to more than just the buffer in using "consciousness," that is, they were also referring to the executive processes. Here is an example where Dehaene is using "workspace" to refer just to the buffer: "This idea is not new, it dates back to the inception of artificial intelligence, when researchers proposed that subsystems would exchange data via a shared 'blackboard,' a common data structure similar to the 'clipboard' in a personal computer. The conscious workspace is the clipboard of the mind" (Dehaene 2014, 166). This must be the sense of "workspace" that means buffer. Most other times, though, Dehaene uses "workspace" to mean "workspace system," sometimes using that very phrase (e.g., 244).

Since Levy identifies consciousness with the global workspace—for example, he refers to "the GWS [global workspace] account of consciousness" (Levy 2014, 47)—we should expect to find the same ambiguity in his uses of "consciousness." Here is an example where he uses "consciousness" to mean "buffer": "[C]onsciousness makes information available for the rational control of behavior by making it available for use in deliberation and reflection, where these are understood as effortful and controlled processes" (Levy 2014, 63). On this understanding of consciousness, the executive processes, which are effortful and controlled capacities that achieve deliberation and reflection, are outside of the scope of what "consciousness" refers to. Consciousness simply contains certain information in a way that makes it available to these other, consumer processes.

Here is a second example where Levy is using "global workspace" to refer to the buffer, excluding the processes that operate on the buffer: "[T]here is a great deal of evidence for the claim that the GNWS [global neuronal workspace] is not merely a set of simultaneous states but a single coherent state" (Levy 2014, 50). The coherence he is referring to is the coherence that exists between the different parts of a (phenomenally) conscious state. As we noted above, they are all bound into a single unified state, produced by processes including thalamocortical oscillations. Levy cannot be lumping the workspace buffer together with the executive processes under the phrase "single coherent state" because the executive processes are not unifying with the conscious state in any such way; they are operating on a separate, unified conscious state. Once the conscious state is constructed, it can then interact with the executive processes and other processes. But this interaction, which apparently occurs via the brain's system of long-range white-matter fasciculi, involves different processes, operating over longer distances, from the processes that unify the conscious state.

In contrast, here is an example where Levy wants to include the executive processes in the global workspace: "It is, however, one thing to show more of the brain, and especially those areas of the brain involved in executive functions (like frontal and prefrontal cortices) are engaged by conscious processing than by nonconscious, and another to show that these activated areas form a GNWS whereby information is made accessible to many distinct mechanisms" (Levy 2014, 50). Here is another example where Levy uses "consciousness" to refer more broadly to the larger workspace system, or at least to the executive processes, a crucial broadcast recipient: "I will argue that consciousness is needed for flexibility of response: for initiation of behaviors that are situation-appropriate, and for the online modulation (or inhibition) of behavior" (Levy 2014, 70). This second, larger sense of "consciousness" has it performing what sound like executive functions. The tasks Levy needs consciousness to perform are not tasks a workspace in the sense of a buffer alone can accomplish. Rather, they are tasks accomplished by the executive processes operating on the buffer, that is, tasks accomplished by the larger workspace system. Indeed, inhibition and initiation of behaviors are two core executive processes, as we saw in chapter 2.[22] In this regard, notice that Levy speaks of "conscious processing" rather than consciousness itself, or conscious states themselves (Levy 2014, 50). Conscious processing, which we would allow is roughly the same as conscious cognition, is accomplished by the entire workspace system, not by the process of consciousness itself.

Levy responds directly to the localist views, claiming they are speaking only about phenomenal consciousness (Levy 2014, 54–5).[23] He argues, though, that even the localist theorists admit that "there must be global activation of networks extending to fronto-parietal areas if there is to be access to and reportability of the stimulus." Thus, argues Levy, these localist theorists "are not dissenters from the integration consensus, as far as awareness is concerned; they are proponents of it" (55). As we noted above, this might be true of awareness, but it is not true of consciousness. Awareness, on our account, is not a type of consciousness, but a causal extension from consciousness to the executive processes. In addition, someone could agree that the frontal lobes must be involved if there is to be reportability without agreeing to the idea that the conscious state *itself* extends to those fronto-parietal (executive) areas. Anything appearing in the workspace buffer can be accessed and reported about once the executive processes are engaged.

The confusion about what exactly in the brain is to be identified with consciousness arises again where Levy discusses Crick and Koch's (1990, 2003) theory that gamma-range oscillations are an NCC (2014, 51). Notice that this theory of consciousness cannot be a theory of consciousness that equates it with a full workspace system since, as we noted, no one has claimed that both the buffer and the executive processes are linked by these oscillations or any such process. Rather, Crick and Koch are claiming that the contents of the *buffer* are linked by gamma-range oscillations. Levy states that "several researchers, most notably Christof Koch (2004), have proposed that consciousness arises from the synchronization of oscillations which bind together perceptual information into a single representation" (Levy 2014, 53–4). This is correct, but note that what is being bound is a "single representation," that is, the conscious state, the contents of the buffer, not the entire workspace system. On our approach, it makes perfect sense to equate consciousness with the workspace buffer but makes little sense to equate it with the entire workspace system.

The executive processes are not bound into the conscious state since the coherent, phase-locked oscillations that bind the conscious state do not extend to them. Since these oscillations can occur on a small spatial scale, Koch's is a localist view. However, Levy lumps the two opposing theory classes together, the localist approach of Crick and Koch, and the globalist approach of the global workspace theorists: "These theories [Crick and Koch's theory and the global workspace theory, as well as that of Tononi]," says Levy, "advance similar and overlapping ideas and marshal overlapping sets of evidence, as their proponents recognize. Their proponents disagree about many things—for instance, when consciousness arises: from early brain-scale synchronization, or from processes downstream from this state—but they agree that consciousness arises from mechanisms that integrate the activity of widely distributed, distinct, and dissociable mechanisms" (Levy 2014, 53–4). These are substantial disagreements, however, between two large theory classes. As we noted in the previous section, Koch does not hold that conscious states themselves are widely distributed.

Distinguishing Consciousness from Control

We are arguing that executive control is the crucial mental component needed for responsibility, not consciousness, as Levy claims. While consciousness is a vexing phenomenon that has eluded science for millennia, the notion of control is not a mystery and is studied by several disciplines, including branches of computer science and engineering. The science of real-world industrial control systems, which these days typically handle the running of certain parts of large factories, is highly developed and not at all mysterious. These systems receive information from various sensors and issue commands to mechanisms that alter different aspects of the factory, such as the speed of its production lines. They use computer programs with commands that might translate as "When a certain part on the assembly line is measured to be out of tolerance, stop the line and activate a siren." Such systems often work with a full and detailed model of the factory itself, analogous to the workspace buffer.

Unlike the well-understood topic of control, consciousness itself remains mysterious. The quest to understand consciousness is a quest to understand our own experience. Why is the object of our experience the way it is? Why does it contain colors, sounds, smells, feels, and emotions? Why are they all together in one grand play? There is nothing about wavelengths of light, vibrations in the air, chemical gradients, and mechanical impingements on our skin that must produce the sights, sounds, smells, and feels we experience. Indeed, there are people called synesthetes in whom the conscious sensations and the sensory modalities are crisscrossed. Vibrations in the air produce colors in their brains, or even tastes, rather than sounds. The core mysteries of consciousness are present in phenomenal consciousness.

One important consequence of the arguments and evidence presented by Koch and his collaborators is that they may show that phenomenal consciousness is the only kind of consciousness there is. Access consciousness is just a phenomenon that occurs when a conscious state is broadcast and currently engaged in causal relations with the executive processes and the other broadcast recipients. Similarly, awareness—cases in which someone is aware of something—are complex mental states in which a phenomenal state is causally engaged with the executive process of attention. Consciousness as a natural kind is a local process, specified at the level of clusters of cortical neurons together with supporting subcortical neurons,[24] not a global process spanning prefrontal and posterior cortices.

While they are separate processes, consciousness and the executive processes work closely together. Like a needle and its thread, they are separate things, but their spheres of function vitally include one another. We believe that Koch (2004) is correct in stating that the primary function of conscious states is to provide an "executive summary" to the prefrontal lobes, a highly processed version of sensory, mnemonic, and emotional information presented to the brain's highest-level control centers for use in producing effective actions. This executive summary is primarily constructed for the executive processes on our approach, although other parts of the brain link to it. On the other hand, defining consciousness as intrinsically including the executive functions, as the globalists do, is like defining needles as necessarily threaded. If you do that, a "needle" without thread is not a needle, on your theory. Similarly, a "conscious state" without active prefrontal connections is not a conscious state, according to the globalists (which, as we will see in chapter 7, accounts for their difficulty in classifying dream states and states of "autopilot driving" as conscious states). According to our view, it's still a needle, and a threaded needle is not a different type of needle. It is a simply a needle that is functionally interacting with its thread. Rather than being a type of consciousness, access consciousness is a use that consciousness is put to. It occurs when the cognitive control network has active causal links with the hot zones containing the conscious state.[25]

The concept of *awareness* works better for Levy, in that it seems to lack a strong "phenomenal" sense, something that must be related to the fact that there are no "aware states" in the way there are conscious states. Awareness also typically occurs in the transitive sense, "Jan is aware of x," which also suits Levy's theory better. Awareness and consciousness are different. Awareness is a state of the workspace system in which an executive process is causally engaged with a conscious state, on our approach. Consciousness is a property of the workspace buffer. A second problem with Levy's emphasis on the "x is conscious of y" sense—the transitive sense—is that the "x" term is actually doing a lot of the explanatory work of such claims. More on this is coming up in chapter 6.

It can happen in science that two competing theories actually have different definitions of what it is they are competing to explain. For instance, Einstein derived a theory of mass designed to supplant Newton's theory, but Einstein's theory also *redefined* mass with a new formula containing the speed of light (Field 1973). The globalist and localist theories are competing theories of something called "consciousness," but they call different things by that name and assign different concepts to it. When a competition of this sort occurs, there is competition not only involving evidence for the two theories, but competition among the concepts themselves. As part of determining which theory is best, we should ask which of the two concepts successfully delineates the phenomena in which we are interested in a way that produces the most useful and powerful theory.

Since it looks as if "phenomenal consciousness" and "access consciousness" are designating different things, we may be in a situation where future science is going to decide which one of them will get the privilege of being known as the process of consciousness. The other theory will be seen as employing "consciousness" in some sort of derivative and possibly even misleading sense of the word, which some future scientists might argue ought to be eliminated. We think data such as that mustered by Koch et al. militate strongly toward phenomenal consciousness. Perhaps sensing that they are engaged in a dispute in which there is no workable middle ground, those on the other side have declared that "all consciousness is access consciousness" (Dennett 2001, Baars 1995).[26]

Once we have distinguished consciousness from the executive processes, if we are right that the executive processes are doing the real work of making us responsible, it looks as if there could be a being with phenomenal consciousness and no agency or responsibility at all. Conversely, there could be a being, some sort of alien perhaps, without consciousness but possessing a cognitive control system that allowed it to be responsible.[27] Once an empirical criterion for the presence of consciousness is refined, this opens up the possibility that all sorts of creatures with conscious states could exist that lacked the brain structures—the minimal working set of executive processes—to be held responsible. If it is true that bees, for instance, have conscious states, as has recently been argued (Barron and Klein 2016), one doubts that we will ever hold them responsible for stinging someone. Levy would argue that more is needed than just consciousness, defined in our way as phenomenal consciousness. One must have the ability to be conscious of the morally significant features of an act. However, what exactly does this consist of? Perhaps animals cannot do this but juveniles can. Juveniles can certainly evince reactive attitudes. Yet we do not hold them fully responsible because their frontal lobes (and hence their executive processes) are not yet developed.

We suspect that the reason why Levy's notion of conscious information's being poised for report appears to bear on responsibility is that such information is either a current focus of attention, or recently was. This may indicate that the explanatory power of the idea of being poised for reporting is actually coming from the executive process of top-down attention. If reportability requires a degree of attention, then reportability would seem to entail executive contact, which is what the above-cited studies that compared report versus no-report conditions show, on our interpretation. Reportability is a reasonable criterion for awareness since awareness involves the executive processes, which are capable of organizing and producing the mental events needed, such as attention, for a report to take place. As a criterion of consciousness, however, reportability fails, as the case of peripheral vision as well as the experimental results from the no-report paradigms show. I can't tell you exactly what objects are at the far edges of my visual field, but I do know that my conscious visual field extends out to whatever is out there.

The attempt by Levy to use access consciousness to refer to both the workspace and the workers splits into two wings, depending on whether "access" means accessible, that is, the executive processes could access the workspace, or whether it means "accessed," that is, the executive functions have accessed the workspace, in that they currently are causally engaged with it. The version according to which conscious states must actually be accessed by the executive processes encourages the notion that it might actually be the executive processes that are doing the real work with regard to responsibility. Levy responds to this by providing an example of unconscious racial stereotyping as a way to show that information not accessed by the executive still has effects on cognitive processes over and above noncognitive processing.[28] We agree that this can happen. But notice that the example does not work against our claim that mental states need to engage with the executive processes in certain ways before a person is responsible. There is a serious question about our responsibility for apparent racist acts driven by unconscious bias. Our position on this is that the person's culpability for racist acts goes up gradually with each act. This is because, at some point, the person becomes guilty of an executive failure: failure to inhibit his or her current way of acting when he or she should (see the second clause of our criterion 3, in chapter 3) and engage in the cognition required to correct the problem by considering different strategies, devising a plan, and so on.

Conclusion

We agree with Levy that events in which the agent is consciously considering the moral significance of a planned action are vitally important to the mental component of responsibility (for positive, synchronic cases at least). These are very large brain events, made up of numerous cortical and subcortical networks and areas, with several different parts, that is, the conscious state and the set of recipient processes, including the executive processes. We differ from Levy, then, on the issue of what portions of those events are most crucial to responsibility.

Ultimately, responsibility has little to do with the actual brain phenomenon of consciousness, the coherent, phase-locked oscillations behind states of phenomenal consciousness. The executive processes do the most important work in the workspace system, not the process of consciousness. They control both the process of conscious deliberation and the creation of actions based on it. Depending on how we interpret "workspace," we get different versions of the global workspace theory, with different conceptions of what consciousness is. The globalist approach sometimes refers both to the workspace buffer and to the processes that operate on it when speaking of consciousness. The localist theory counters with its concept of consciousness and its interpretation of that data.

Levy notes that "the degree of accessibility of information seems to correlate (roughly) with the degree of moral responsibility of the agent for failing to utilize it" (Levy 2014, 32). We would suggest, though, that the reason why consciousness of a feature increases our moral responsibility is not because of consciousness itself, but rather because that sort of consciousness normally signals executive engagement. When we say Sam was aware that there were people in the house he set on fire, we are claiming that there was executive engagement. We are saying that Sam's cognitive control network entered into the right sort of causal activity with his conscious representation of the fact that there were people in the house.

In his concluding remarks, Levy says, "In this book I have argued that consciousness of the moral significance of our actions—those features in virtue of which our actions have their moral valence—is a necessary condition for direct moral responsibility" (Levy 2014, 131). There is still a gap, though, between being conscious of moral significance, on the one hand, and the ability to respond flexibly, rationally, and intelligently to that awareness of moral valence on the other. Consciousness itself does not provide the flexibility. Rather, the processes that consciousness broadcasts to, especially the set of executive processes, provide the flexibility. Consciousness of the moral significance of an action you are considering does you no good if you are unable to control and regulate your behaviors in a way that respects that moral significance. If you are unable to plan actions in a way that respects that moral significance, or unable to inhibit contemplated actions that fail to respect that significance, or unable to monitor your actions as they unfold in order to alter them when they fail to respect that moral significance, then you are not a responsible being.

6 Judgments and Claims of Responsibility

In this chapter, we will argue that our everyday ways of attributing and talking about responsible acts are broadly consistent with an executive theory of responsibility. In our roles as folk assessors of responsibility, we maintain a type of sensitivity to executive activity when making our judgments, partly via perceiving behavior, but also by using mind-reading capacities, we will argue. This sensitivity also shows itself in the ways we form and assess claims about responsibility. This broad agreement between our theory and folk psychology in turn bodes well for harmonious relations between our executive approach and the legal system, given the traditional closeness between folk psychology and the law.

The legal system uses the language of folk psychology, augmented by technical terms. It frequently provides more specific definitions for the terms of folk psychology—for example, in defining the concepts *purposefully* and *knowingly*—as we discussed in chapter 4 and will readdress in chapter 10. Legal procedure allows the two opposing sides, as well as the judge, to force clarity of folk-psychological claims. One reason why the two parties might need to clarify language is a chronic imprecision in our action verbs: We tend to use the same verb for both intentional and nonintentional versions of an action. However, intent is crucial to the law. If all the witness says, for instance, is that Harry *killed* Sam, the next question on everyone's mind is whether Harry did it on purpose. We experience cognitive dissonance as we vacillate between the intentional and nonintentional readings of that claim and the mental scenarios they invoke in us. When we imagine these scenarios, emotional responses rise up, our reactive attitudes. Rather than dither between those two options, we demand disambiguation. We need specific information about the mental states of the accused. Did he *plan* the murder? Did he *intend* to shoot the victim? Did he attempt to conceal his act? If so, our reactive emotions are much stronger and more negative, and we will judge the defendant much more harshly.

But can we discern, with a level of exactness fitting for the legal system, when folk-psychological claims about people are true? We think the answer to this question is "yes." We believe the potential precision present in folk psychology is underestimated. Children master a vast set of conventions that govern the uses of our words. As part of learning a native tongue, we learn an intricate and exact set of rules governing the use of our mental concepts, such as *know, see, plan,* and *remember.* We know it is false to claim that a rock is conscious of something. We know that little children can't grasp abstract ideas. We also know a lot about the connections between conscious states and other mental states. We know that certain mental states tend to cause certain other mental states, such as the way that frustration leads to anger. In what follows, we will argue that folk psychology has unappreciated precision when it comes to claims about responsibility.

Assessing Claims about Responsibility

In order to know whether someone is truly responsible for something, we need to know about the person's mind, according to our account and most others. But, a familiar objection goes, how much can we really know about the mental states of others? Aren't we stuck making inferences from their behavior? We are not actually speaking about people's minds or brains when we make these statements about responsibility; we are merely talking about their current, and perhaps future, behavior, or their behavioral dispositions, goes the behaviorist's line, still popular among philosophers. It turns out, however, that our standard ways of talking about conscious states and other mental states go much deeper than behavior.

One clear way to see this is via a curious linguistic phenomenon known as referential opacity. Here is an example: Jen has just started reading the work of the writer Mark Twain, and she loves it. However, she doesn't know that he was also called by his given name, "Sam Clemens." But as it happens, Jen works with a man named Sam Clemens, who she knows well. So, at this moment, while it is true to say, "Jen loves the work of Mark Twain," it sounds wrong, or false, to say, "Jen loves the work of Sam Clemens." One obvious reason for the wrongness of the second sentence is that we know that Jen will not assent to it. But our knowledge goes deeper than mere behavior, such as Jen's unwillingness to assent. We know the *reason* why Jen will not assent to it: Her attitudes do not represent the writer as Sam Clemens. Normally, this phenomenon is described using the concept of belief. For example, we can say, "Jen believes that Mark Twain is a writer," but it sounds false to say, "Jen believes that Sam Clemens is a writer." The part of the sentence that follows the "believes that" clause is said to be "referentially opaque" because we are blocked from making substitutions in the wording in that sentence, even when we substitute words with the same referent.[1] But, the "loves" case shows that this phenomenon does not require a statement with a "that" clause to show itself (e.g., Jen loves Sam). Referential opacity is present for virtually every mental verb, including knowing, understanding, desiring, planning, loving, hating, seeing, hearing, and on and on. When we apply them to someone, we are sensitive to *how* that person represents the situation, and we often try to represent the situation the way that person would, typically by using the language that person would use.

Interestingly enough, our attributions of responsibility also show opacity effects. Consider the case of Fred, the impatient driver. Fred is making his daily drive to work. It is a tough drive since it involves a lot of traffic lights, difficult merges and other maneuvers, and ultimately lots of patience on Fred's part since the traffic is always heavy at this time. Fred's patience fails him on this particular morning, and after a driver in a gold car in front of him fails to move quickly enough after a light turns green, he maneuvers around the car, and as he passes it, raises his left hand and extends the middle digit, giving that near-universal sign of disdain—"the finger"—to the driver of the gold car, whom he only perceives as a man wearing a hat. But when Fred gets to work, he is summoned to his boss's office and unceremoniously fired: The man in the gold car was his boss, who recognized Fred during their encounter. Now, when Fred's coworkers need to explain why he is packing up his desk, what should they say? Should they say, "Fred got fired because he gave the boss the finger"? But he did not do so knowingly. Thus, they should add, "even though he did so unknowingly," if they want to be accurate. We need to know Fred's representation and understanding of what he did to understand his responsibility. Is it correct to say that Fred is responsible for giving the finger to the boss? There is an opacity blockage that makes the sentence sound wrong. This occurs because the mental level of our analysis of the sentence fails, because we know, in some inexplicit way, that Fred's executive processes were not interacting with his normal mental representation of his boss at the crucial moment, we will argue. Fred did not intend to direct an action at his boss. We judge that this failure at the mental or internal level of the ascription makes Fred significantly less culpable for his behavior toward his boss.[2] In another scenario, Fred recognizes his boss and intends to wave to him. But, alas, he forgets that he broke the middle finger on his left hand which is now in a splint and stands well clear of the other fingers—clear enough for the boss to take it as an insult, and again Fred is fired. However, is Fred responsible for giving the finger to his boss? We balk here too. Fred did not intend to direct the finger at his boss. That is not how he represented what he was doing. Thus, it appears that the ascription must depict both Fred's act and the target of his act, the victim, correctly in order to satisfy our conception of responsibility. Our folk concept of responsibility is infused with mentality, and the phenomenon of referential opacity is a by-product of this, we suggest.

An obvious first hypothesis as to why we would go through all the trouble to do this, to track the interactions between the executive processes and the person's stock of representations, is because of its predictive value. Prediction of a person's behavior is much more accurate when coupled with a knowledge of how that person represents his or her situation. To borrow an example from fiction, I can predict how Lois Lane will behave around Clark Kent (aloof) and how she will act around Superman (starstruck) once I know that she has two mental representations of the same person, and which attitudes are associated with those representations.

There is a long-running and subtle debate on the issue of whether attributions of attitudes are actually made false by this failure to describe the target's mental states, or whether the utterance is inappropriate, odd-sounding, or just less informative.[3] If the failed mental requirements actually affect the truth value of the attribution, they are said to be semantic features of the attribution; in the other cases, they are likely pragmatic features of how we assess the appropriateness of a claim in a given context. Our example above: If changing "Mark Twain" to "Sam Clemens" turned a true sentence into a false one, this indicates the presence of a semantic role for the names. We suggest that we are indeed talking about semantic rather than pragmatic factors in cases of responsibility for serious moral or legal violations because of the way that the high stakes of the situation force a focus on the truth of the attributions. But in our analysis, we will bypass this issue by speaking of attributions as being complete and correct—complete, in that they refer to the right entities both at the mental level and the external level of behavior, and correct, in that these entities are related as claimed, typically by the verb of the attribution. We retain an acute sensitivity to both these levels, and that produces the phenomenon of opacity.

We are sensitive to the ways in which people consciously represent the objects they have attitudes toward because we are tacitly aware that those are the states that are interacting with the executive processes, we would argue. And we tacitly "know" that if we want to understand and predict their behavior accurately, we need to simulate their minds accurately. Our folk attributions appear to contain a structure that works to organize our knowledge about mental states and allow us to apply our concepts of such mental states in consistent and effective ways. Folk psychology represents responsible actions in the form of a structured representation which has two parts, or levels, in our approach. One typically represents people causally interacting with each other or with medium-sized objects (we'll call this level the external level) while the other typically includes the cognitive control network—the executive processes—causally interacting with conscious, representational states (we'll call this the mental level).

For example, suppose Sam is standing several paces from a tree, looking at it. There is a squirrel climbing on the trunk of the tree, and Sam's eyes are directed at it, but he hasn't actually detected the squirrel yet, because it is too well camouflaged. Is it complete and correct to say that Sam sees the squirrel, based on what we know so far? No, because as we all know, people can fail to see things even when their eyes are directed at them. The statement, that Sam sees the squirrel, fails at the mental level since it fails to characterize the interactions of his executive processes with his mental representations correctly. If Sam did see the squirrel, his executive process of attention would be causally connected to his visual representation of the squirrel in certain distinct ways. As materialists, we regard both levels of the structure as consisting of physical objects engaged in certain causal interactions. What the mental level of an ascription does is ensure that the cognitive level is in order in the target person. This means that there are adequate representations, connected in the right way, to the right executive processes.[4]

But, an objection goes, in what sense are we *referring* to the target person's executive processes and mental representations, since most people are not aware that they are making such references? When we say, "It is raining," we leave out any mention of where exactly it is raining, and when it is raining, because we all assume that we are claiming that it is raining *here* and *now*. The place in which it is raining is thus an *unarticulated constituent* of the proposition[5] we asserted when we claimed it is raining, according to philosophers of language Mark Crimmins and John Perry (see Crimmins 1992, Crimmins and Perry 1989). In the language they use, we *tacitly refer* to the place (as well as to the time). Based on the way we respond to the story about Jen and Mark Twain, it looks like we have a way of representing how other people represent things. We know that Jen represents Mark Twain with the name "Mark Twain" and not with the name "Sam Clemens." Crimmins and Perry argue that we are tacitly referring to Jen's mental representation of Mark Twain when we attribute beliefs to her (and we are generalizing this phenomenon to include other types of attitudes).[6]

So, what brain processes are achieving the references and the tacit references? In most cases, the (nontacit) references we explicitly make, that correspond to the terms in the sentence, are achieved via our normal stock of representations. If I see Jill shoot Jack, I form the belief that Jill shot Jack by forming some sort of link among my representations of Jack, Jill, and shooting. As for the tacit references, what makes and "hears" or "understands" them? We think that a good candidate here is the brain's default mode network, as described in chapter 2, which has mind-reading as one of its functions. We know of no direct empirical support for this claim, but there are studies showing default mode activity as subjects are hearing sentences.[7]

The set of executive processes is not unique in having access to a person's present conscious states, according to most versions of the global workspace theory of consciousness (which recall from chapter 5 that we can endorse provided consciousness is distinguished from executive processes correctly); limbic areas, mnemonic areas, and other areas are also receiving the broadcast of the conscious state. Nevertheless, the executive processes are the only ones on this list that have the sort of causal contact with conscious states that they do. These causal interactions give the executive functions a sort of active presence in consciousness—a self-like quality, we would suggest.[8] What has historically passed as the subject of conscious experience, sometimes referred to as "the self" (as in Locke, Hume, and James), or "the ego" (as in Descartes) may simply be the brain's control system, the cognitive control network, or the set of executive processes. A survey of historical philosophical writings on the self, or the ego, shows that the list of functions assigned to it—willing, understanding, deciding, and so on—is simply a list of executive functions. As we noted in the previous chapter, the executive functions are not part of the mystery of consciousness because they are separate from consciousness, and never appear in consciousness. Again, the history of philosophy is consistent with this. The Scottish philosopher David Hume expressed this centuries ago when he rejected the idea that we have conscious awareness of a self—something that performs actions in the mind: "For my part, when I enter most intimately into what I call myself, I always stumble on some particular perception or other, of heat or cold, light or shade, love or hatred, pain or pleasure. I never can catch myself at any time without a perception, and never can observe any thing but the perception" (1739, book I, part IV, section VI).

The executive functions supervise the workspace system; they conduct the planning and decision-making processes by using the brain's resources: searching the declarative memory store and verifying any putative memories that appear in consciousness; engaging with the default mode network in order to evaluate hypothetical claims, claims about the future, and claims about other people; employing working memory to solve problems, resolve conflicts, and make decisions. As we noted in the introduction, we seem to acknowledge the important roles the executive functions play in our minds by using that special word "I" to (tacitly) describe the different executive activities. Intending, planning, deciding, attending, and inhibiting are all things that *I* do. "I" at the external level refers to the person who utters it, but at the mental level, "I" can refer to an executive process. There is another interesting covariation between what the executive processes can do and what "we" can do: What we can do in the cognitive realm corresponds to what the cognitive control network can do. Consider the following examples: "I didn't perform that automatism during my seizure," "I (the color-blind person) cannot make myself see the red light as green," "I cannot detect the visual blind spot directly," or "I don't have access to lower levels of perceptual processing." In general, the parts of perceptual processing that *we* are aware of correspond to those parts that the cognitive control network has causal access to.

There is a further covariance that relates more closely to our view: We tend to accept the actions of the executive processes as our actions: We say, "I stopped myself from doing x," "I planned to do x," "I switched my attention to x," and so on. And we tend to accept responsibility for what the executive processes do when we acknowledge their processes as our actions. For example, I acknowledge executive activity directed at planning a vacation as my planning (and I will accept responsibility for any inadequacies in it). In contrast, there are clear cases where we do not feel responsible for what merely appears in consciousness. If the thought occurs to someone to grab the cash out of the register while waiting for a clerk to make change at a convenience store, the person is not responsible for that; silly thoughts like that happen all the time. But if the person begins forming a real intention to do it and planning an escape route, that person's responsibility level rises considerably. When conscious states merely occur to us, we describe ourselves passively: "An idea popped into my mind" or "This song keeps playing in my head." All of this is, of course, fogged up by our tendencies to acknowledge ownership of mental acts consistent with our beliefs about ourselves while denying ownership of mental acts that are inconsistent with this self-representation. Most of us employ folk psychology with full awareness of this problem, however, and we often make adjustments to our mental models of people based on this.

On the negligence/failure side, we tend to accept and assign responsibility for failure to employ our executive processes in certain crucial contexts where care is needed. For example, consider the school bus driver's failure to make a final check of the bus before heading back to the garage, locking a sixth grader on the bus overnight. Even though the bus driver makes that check routinely, in this case it needs to be a *monitored* routine. Top-down attention must be operating to make sure the action runs its course properly, which means that the brain networks that accomplish top-down attention are performing their monitoring function. We often monitor routine activity when we need to. When the normal drive to work gets hazardous, because of heavy traffic, snow and ice, or heavy rain, we pay extra attention. And we insist that others do also, on pain of powerful negative reactive emotions from us, not to mention the legal consequences we have devised for such failures. We will discuss a case of criminal negligence at length in chapter 7. When the bus driver fails to perform that check, there is an executive failure, and we will hold the bus driver accountable. Again, we sometimes cheat here, being much quicker to assign responsibility to others than to accept it ourselves. However, most all of us know about this phenomenon, and we account for it when we perceive and understand what people say and do.

Our natural reflex thus seems to be to "identify with" executive processes, but this appears to be modifiable.[9] For example, inhibition is standardly included in the list of executive processes. It is apparently accomplished several different ways by different brain processes, depending on the sort of action or intention that is being inhibited. Normally, when an inhibition occurs, we might describe that by saying something such as, "I stopped myself from doing x." Some people with OCD, however, are taught to separate their sense of identity from a chronically overactive prefrontal inhibitory process that stops them from doing anything until, for example, their hands are clean or the locks are checked. They learn to "disown" the prefrontal activity: "*I'm* not the one who wants to wash my hands; it's just some overactive brain process." "It's not me; it's my OCD" goes the mantra sufferers are urged to invoke when the inhibitory activity grows strong.[10] Not thinking of this activity as their own seems to substantially reduce its power to hijack the brain's normal function.

Recall from chapter 3 that one criterion in Fischer and Ravizza's account of responsibility is that the person must take responsibility for his or her reasons-responsive mechanism. Our approach provides a concrete, empirical way to study the phenomena involved when we take responsibility for our actions—by looking to executive activity, as we would phrase it in the language of our theory. Conversely, disowning one's own executive processes can be a sign of irrationality. "There is thinking going on in me, but it's not my thinking" sounds clinical, perhaps schizophrenic. In this regard, the ancient philosopher Socrates is rare in making his daemon, a conscience-like inhibitory voice that keeps him from committing wrong actions, something apart from himself. The case of Socrates appears to show, though, that a person can disown one executive process and still be quite rational. But schizophrenics, some of whom seem to have lost control of their own minds (and hence bodies), are considered irrational primarily because of their pervasive executive dysfunction, we suggest. We will take up the specific case of schizophrenics in chapter 9.

Just as we tacitly refer to our own executive processes when we say "I" in self-attributing mental events, we also tacitly refer to the executive processes of other people, typically by using their given names, it appears. If we learn that the relevant executive activity is absent, we will revoke an earlier use of a mental verb and/or a use of a person's name.[11] For example, as we will see in the next chapter, we reject the claim that Kenneth Parks intentionally killed his mother-in-law when we learn he suffers from a sleep disorder and was actually asleep and in the middle of a type of dream when he stabbed her. The necessary executive activity is absent; he neither planned nor intended to kill her. We similarly reject the claim that a drug addict named Ariel wants or desires heroin, even when she drives some distance to purchase it, when we know that Ariel herself hates using it, considers it a horrible habit, wishes to stop, and so on. Ariel's considered desire is that she wants to not want heroin. She explicitly makes plans to end her addiction, but some nights she just finds herself driving toward her fix. We revoke the intentional state claim because one of the important conditions for the completeness and correctness of the claim has not been met. Her heroin habit has not received "executive approval."

The process by which conscious states are exposed to the set of executive processes gives the executive processes time to perform numerous checks on them. This is clearly illustrated in those cases where executive functions work to verify putative memories. For example, it seems to someone that she remembers being at an Obama campaign rally in 2012, but upon further thought, as well as checking this alleged memory against other memories and the Internet, she realizes that it was actually a rally in 2008. Similarly, in the case of planning, conscious states representing certain actions in a context are created: "If I go into the boss's office to ask for a raise, it's going to go like this..." At that point, the executive functions have the opportunity to inhibit that plan from forming into concrete action: "If I just ask for a raise straight out, that will come across as demanding, and I probably won't get it, knowing my boss."

These phenomena indicate that we employ something like a notion of executive "approval"—in these cases approval of a memory and a plan—in taking ownership, as it were, of certain of our conscious thoughts once they have been processed in the ways described. It appears that we also use a tacit notion of approval when we assess other people as being responsible for an action.[12] We hold people responsible for actions that we believe have received their executive approval, or where there ought to have been executive involvement. Note that it does not follow from the fact that someone's executive processes are interacting with his or her representations of harming x that his or her executive processes have approved harm to x, that he or she attaches a low value to the life of x, and so forth. In general, executive activity with regard to harming x does not imply executive approval of harming x. It must be the right sort of executive activity, resulting in final approval. In some cases people are fantasizing about doing something unethical with no intention of actually doing it. (Such people are responsible for the act of fantasizing, if it happens in a repeated and deliberate way, however.) We suspect that actual planning will be easily distinguishable from mere fantasizing by looking at brain activity—specifically, executive activity. Indeed, part of the reward value of fantasizing about doing something one has no intention of actually doing is that it does not involve the pointed, rigorous, and stressful executive activity that making a real plan to act would.

Speaking Naturally

If a responsibility theorist is a materialist/physicalist, the statements that make up his or her theory, and their parts, the noun phrases, adjectives, indexical phrases, and demonstrative phrases, should refer in a straightforward way to physical events, things, and properties. Progress, mostly in science but also in philosophy, has reached a point where stand-ins are unsatisfactory. Responsibility theories may now be clear about which physical things are relevant to responsibility ascriptions. On our account, mental state ascriptions refer to two levels of physical objects. At the external level, they refer to people interacting with other people and things. At the mental level, they refer to the cognitive control network interacting with representations of people, things, and events.

One great benefit of putting things in clear physical terms is that attaching truth values to mental state ascriptions becomes much more straightforward. There are, however, several current physicalist accounts of responsibility that contain terms in their analyses that do not refer clearly to physical things, including Neil Levy's account. Recall from chapter 5 that, according to his theory, if a person is to be held responsible for an action, at some point in the person's past, there was an event in which that person was conscious of the moral significance of that action. However, when we attempt to analyze that event, and that awareness relation, we find that something prevents us from describing the event as a physical event and the relation as a physical relation. Levy says that "the consciousness thesis is the claim that *an agent* must be conscious of (what she takes to be; here 'conscious' is non-factive) the facts concerning her action that play this important role in explaining its moral valence; these are facts that constitute its moral character."[13] And in further specifying this, Levy says, "[T]he kind of consciousness at issue—awareness—is a state with contents of which *the agent* is aware."[14] But here is our question: How do we cash this out in terms of clear physical referents? In particular, to what physical thing does "the agent" refer?

Take a case where Levy's consciousness condition has been met. There is a conscious state in which an agent is aware of the facts that constitute the moral significance of an action. Awareness is a type of physical-causal relation according to both Levy and us, let us assume. However, what are the relata? Suppose we want to trace the causal chain from the consciousness of the moral significance of an action to the action itself. How does the conscious state causally connect with the action? What does it mean to say that it does so via the agent? It seems that between the consciousness of the moral implications and the action is something called the agent. What is this, in clear physical terms?

Levy sometimes speaks of *accessibility* between information and the agent. "The degree of accessibility of information seems to correlate (roughly) with the degree of moral responsibility of the agent for failing to utilize it" (Levy 2014, 32). But how is the metaphor of access to be cashed out here? Presumably it is intended to be a causal notion. For instance, to say that one physiological system, such as the nervous system, has access to another system, such as the immune system, means that there are certain types of causal processes running from the former to the latter (and perhaps also the other way). Levy also uses the notion of *availability*. He says that "information that is available enough to *the agent*, I suggest, is personally available; that is, available to *the person* (rather than merely to some of the consuming systems of the mind)."[15] Thus, information that is available to the agent causally interacts more broadly than merely interacting with some of the consuming systems of the mind. How does this happen? How does the causality spread to the entire person? Further, "[W]hen an agent has such information personally available to her, she is conscious of it," says Levy (2014, 34). Availability is presumably a causal notion. Here, to say that x is available to y means something like y has ready causal access to x, so that availability describes a causal relation between two things. What are the two things causally related in the case in which an agent has information available to him or her?

According to our view, the set of executive processes in one sense stands in for the person, at the mental level. We suspect that, since Levy does not recognize anything like the set of executive processes, he needs something like the person to interact with the relevant conscious states. Levy thus appeals to the person himself or herself, or some sort of metaphorical extension of the concept *person* or *agent*. Our fear is that this form of expression—"the agent," "the person," and so forth—is not naturalizable as it stands, and must be translated into something that is. We suspect that when it is, the executive processes will be found hiding behind such uses of "agent" and "person." It may be that the reason why it sounds plausible to speak of the person or agent as having access to the conscious state is just *because* in referring to the person or agent (in such cases) we are also tacitly referring to the set of executive processes. By including an essential mention of the agent or person in his theory, Levy sneaks in an essential (tacit) mention of the executive processes.

The Competing Views Gain Their Force by Reference to an Executive-Controlled Workspace

Levy sees his consciousness condition as necessary for reasons-responsiveness. He says that "in order to know which agents ought to be sanctioned, when, and how, we need to know whether their behavior was reasons-responsive; consciousness of the reasons for which we act, I shall argue, is needed for reasons-responsiveness" (Levy 2014, xi). However, states of conscious awareness themselves cannot be the building blocks of reasons-responsiveness. If moral information or reasons just appear in consciousness, if there is just one conscious state after another, this does not seem like a good candidate for something like the ability to respond to reasons. The executive processes move cognition along in a productive way and, ideally, in a rational way. Consciousness by itself just sits there, as in the case of meditation (and possibly akinetic mutism). There doesn't seem to be anyone, or anything, in charge in these cases. Broadcasting is important, but the broadcast states need to be processed in certain ways before that person is reasons-responsive.

According to our theory, the history of a person's set of executive functions may matter to his or her responsibility in diachronic cases. But unlike Fischer and Ravizza (1998), we need not rely on long-term observation to determine whether a person has the capacity to be responsible since our approach is directed at the neural mechanisms that produce the reasonable responses. To use another car analogy, I can know that your brakes don't work without needing to see you actually try to stop if I see that you have no brake fluid. Your braking system is in a state in which it is guaranteed to fail. Theoretically, one could do the same thing with something like the process of inhibition in a person. It is possible that the scientists and doctors of the future will not need a record of disinhibited behavior to diagnose an inhibition problem; they will be able to see the actual damage that will inevitably produce disinhibited behavior. Because our theory allows us to describe the mechanisms behind responsibility themselves rather than only their products, we are better able to distinguish the epistemology of the mental components of responsibility—how we know about their presence—from their metaphysical nature—what they actually are.

We see our theory as providing concrete grounds for how and where reasons-responsiveness occurs. Using the executive processes to understand responsibility can shed light on the specific functions in the brain that underpin the reasoning processes necessary to responsibility and the many ways in which they may work or fail. Further, our more sophisticated understanding of reasoning processes—positing not just a "mechanism" that underpins reasons-responsiveness, but distinct and verifiable brain processes that compose the set of executive functions—allows us to better understand the degrees to which persons may be responsible, depending on how engaged these processes are in producing an action.

However, while it seems obvious that the executive functions would play a crucial role in allowing people to reason and make decisions, we realize there are many more details to be worked out. We can make some obvious points, though, about how the executive functions might help make us rational. The general ability to detect conflicts in one's representations of the world, and then work to resolve them, is one of the building blocks of reasoning. A core principle of reasoning is the law of noncontradiction—logical contradiction counts as a type of conflict. We can also detect and resolve other types of conflicts among our system of representations, including conflicts between perception and belief, between semantic memory and autobiographical memory, and so on. Notice also how crucial the abilities to attend and to task switch are for our reasoning processes. The ability to use the resource of working memory is likewise crucial to human reasoning. We use working memory as a workspace (buffer) to manipulate and analyze representations. We also are adept at offloading cognitive work onto the environment itself. For example, the blackboard is vital to a math class: It functions as a sort of external, public working memory for the class and its professor, displaying "memories" which are "forgotten" when it is erased at the end of class. The executive processes play the primary role in achieving human reason, but they are not perfect. They are prone to several flaws, fallacies, and foibles.

We also think our executive approach can inform real-self theories of responsibility. Levy also draws a fascinating connection between his consciousness-based approach and the real-self views. He says that "only when our actions are expressions of our selves can we be appropriately identified with them, such that we can be assessed on their basis" (Levy 2014, x). The brain's ability to broadcast conscious states plays a crucial role in ensuring that actions reflect the agent's real self. Here is how Levy makes the connection between consciousness and the real self: "I argue that only when attitudes that imbue actions with moral significance or which otherwise reflect that significance are *globally broadcast* are those actions expressive of the agent's evaluative agency, because only under those conditions is the moral significance assessed for consistency and conflict with the personal-level attitudes constitutive of the real self" (Levy 2014, viii). We agree with much of this; however, the process by which the person decides just what his or her real attitudes and values are, and detects and resolves inconsistencies, must be conducted and coordinated by the executive functions. They must mitigate all sorts of conflicts among the recipients of the conscious broadcast—for instance, conflicts between emotion and rules or authority.

As the real-self view affirms, the desires that most reflect my real self are not just knee-jerk desires, but *considered* desires. To us, this means they have the executive processes' stamp of approval. The memories that are important to decision-making are not just memories in the form in which they first pop up into consciousness, but edited memories that have been checked against other memories and relevant perceptions, such as consulting a diary or the Internet. What this means is that, in order to truly reflect a person's real self, the memories or desires need to have the proper executive engagement. It appears that this engagement might be either brief or extensive. The executive processes should inhibit contemplated behaviors that are not consistent with that person's firmly held attitudes. However, we believe that people are also responsible for sudden outbursts that might not reflect their characters or their second-order desires, such as hitting a misbehaving child in anger, contrary to the real-self view. Because such acts should have been inhibited by an executive process—because the part of the brain that serves as the self should have been involved in issuing those actions—we treat them as cases of executive failure for which a person is responsible.

Of all those processes privy to the workspace buffer of consciousness, only the executive functions are necessary for responsibility. This claim is supported by cases where one or more of the recipient processes, including the executive processes, is inactive. We then gauge the effect this has on the person's level of responsibility for actions performed on the basis of that mental activity. First, consider a case in which the executive functions alone are missing from the workspace system: A five-year-old has long-term memory; is able to attach values to objects, people, and events represented by the workspace buffer; and seems to possess all the other broadcast recipients but is not responsible because his executive functions are not sufficiently developed. Children can be aware of the moral valence of an intended action, but we do not hold them responsible because there is too high a chance they do not possess a minimal working set of executive functions. They may lack inhibitions, or they may lack the advanced ability that most adults have to able to employ "gut" assessments of risk in their cognition (as did Damasio's [1994] patient).

Second, when memory alone is missing from the workspace system, people are still held responsible. Suppose an amnesiac commits a crime. She sees someone on the subway platform whose face she doesn't like and pushes him into the path of an oncoming train. Her amnesia offers no excuse for what she did.

Conclusion

When Levy says the person must be conscious of the morally significant features of an act, we believe this contains two errors. First, it is executive engagement rather than any relation to consciousness that does the real work in giving us the capacity to be responsible. The two entities that need to interact to fulfill the crucial mental criteria for responsibility are the executive processes and representations of the intended act and its harm, not the person and the morally significant features of the intended act. Second, Levy is able to gain the explanatory benefits of using the executive processes by using the concept of the person or agent, which tacitly refers to the executive processes. But in Levy's case, this produces a theory that cannot be connected to science until we know what "the agent" or "the person" refers to, as Levy employs them. It looks to us that Levy is using "the person" or "the agent" to refer to both the full biological being and to a part of the mind, the part that has access to the conscious state, that is, the executive processes, since they are what actually interact with conscious states in ways that Levy says the person does.

When we look at the case studies in the next chapter, we will see that several of the existing puzzle cases of responsibility appear to involve a failure of a claim to tacitly refer to the executive processes. It is wrong to say, for instance, that the unwilling addict wants heroin because this desire has not received executive approval. "*I* don't want heroin; my body needs it," he might say, and rightly so. As the examples indicate, our ability to detect failure of tacit reference to the executive processes manifests itself as a hesitancy or refusal to use the normal ways of referring to a person who is doing something fully intentionally (i.e., with executive approval). Hence the italics in the above examples—there is something wrong with using the person's name (in the third person) or "I" (in the first person). In attributing mental processes, we use "I" when the executive processes played their appropriate role. From the third-person point of view, the same can be said of our uses of the person's name. Part of the error made by the globalist theorists of consciousness in including the executive processes as part of consciousness is that they did not notice that when they used the "Jan is conscious of x" construction, "Jan" is also (tacitly) referring to Jan's executive functions. This helps to explain why the globalists failed to notice they were wrapping the "subject" of consciousness in with consciousness—the "x is conscious of y" construction (tacitly) refers to them both.

7 Responsible Lives, Responsible Acts

One way to test our account of responsibility against its competitors is to look at some of the existing cases that have proven difficult for theories of responsibility to resolve. In this chapter we will apply our account to several difficult cases of responsibility, including negligence, REM behavior disorder, unwilling addiction, alcoholic blackout, and flow states. We will go into some detail in the first two of these cases since they provide another chance to compare our theory to Neil Levy's consciousness-based approach. Levy says the crucial ingredient is a certain type of conscious state, but we say responsibility depends upon the role the executive processes play in action. If we can find cases where one of these is present but not the other, we can better distinguish the two theories. How well does the presence or absence of consciousness or executive activity correlate with the presence or absence of responsibility? Which one "tracks" responsibility better?

Is Consciousness Required? The Case of Negligence

In prior writings, one of us discussed the real-life example of Bert (not his real name), who was a defendant in a child neglect case (Sifferd 2016a). Bert had custody of his kids every other weekend, although he often tried to switch days with his ex-girlfriend and sometimes failed to take the kids when he was scheduled to. His children were ages six and eight and were latchkey kids, meaning they let themselves into his apartment on the days they were scheduled to stay with Bert, to wait for him to return home from work. One Friday when he was scheduled to have his kids, however, Bert instead went straight from his workplace to the airport and caught a flight to Las Vegas. There was no land-based phone line working in Bert's apartment and very little food. The children, with no means to contact either parent, stayed there alone until Monday morning, when they returned to school hungry and dirty. The children's teachers figured out what had happened and called the Department of Child and Family Services, which in turn called the police. Bert was arrested for child neglect.

Bert argued that he had honestly forgotten that it was his weekend to care for the children because he was confused about his parenting schedule and thus should not be responsible for their neglect.[1] Statutes covering criminal child neglect, endangerment, and abandonment in the various U.S. states reflect an extremely wide range of mens rea, or mental state, requirements. Nebraska, for example, requires mere negligence, and Maine requires recklessness, but Mississippi requires the standard of "willfully." However, the majority of state statutes do not specify any mens rea and merely prescribe liability for failure to provide necessary or proper care (Pimentel 2012). In Illinois, the state where Bert was arrested, the statute that defines child neglect stipulates that a person is guilty of child abandonment if the person knowingly leaves a child under twelve alone for 24 hours or more. Bert stated he hadn't knowingly abandoned his kids because he wasn't aware it was his weekend. The court, however, found Bert guilty of child abandonment and sentenced him to probation.

We believe most people would hold Bert responsible for the lapse that left his young children alone and hungry, at least to some extent. It may be that Bert honestly forgot it was his weekend to care for his kids, but he is responsible for the harm caused to his children because he is responsible for the forgetting. Human beings regularly hold each other responsible for lapses; big and small, consequential and inconsequential. If my friend forgets to call and make reservations for our planned dinner, I am likely to feel she is blameworthy and to feel justified in expressing my annoyance; we very often hold each other responsible for forgetting birthdays and anniversaries; we feel justifiably upset when our partners forget our plans to meet after work, or when a co-parent forgets to buy something necessary for a child's school project; and we blame a family member who forgets to repay a small loan. Certainly, in some cases we feel only a little bit of moral anger in response to a lapse, but in other cases we might be outraged. Often, the more severe the foreseeable consequences of a lapse, the more moral anger or blame we will feel toward a person for his or her lapse.

On what grounds do we blame others for lapses? There is a clue in the fact that the folk treat lapses that cause harm differently than they treat accidents that cause harm. If my brother trips over a snag in the tassels of a carpet and breaks an heirloom vase so that we can no longer use it to give flowers to our mother on Mother's Day, I may be annoyed but remind myself that it isn't fair to blame him. After all, it isn't his fault there was a snag in the rug—he couldn't have been expected to see the snag, and he was exercising a reasonable level of care when he was carrying the vase. However, if I ask my brother to pick up the vase from the shop where it is being fixed, and he forgets, so that we cannot give the vase to our mother, I will blame him and am likely to express my disappointment. The difference between the two cases seems to be exactly that it was within my brother's control to pick up the vase on time, and if he had been exercising reasonable care, he would have remembered to pick it up. By contrast, it was not within his control to avoid tripping over the snag in the rug—he could not have been expected to survey the rug for snags before he stepped forward with the vase; he was exercising reasonable care and yet still tripped. Blaming my brother for forgetting to pick up the vase seems appropriate under the circumstances, and expressing some anger or blame seems proportional to his lapse (assuming he isn't hypersensitive or hard on himself). Interestingly, my expression of blame may also serve the forward-looking aim of making his future promises to me more salient.

Note, however, if I blame my brother for tripping over the rug and breaking the vase, my brother is likely to call my reaction into question by stating that the tripping wasn't his fault (whereas he may respond to my anger at his memory lapse with expressions of guilt). Blame in the tripping/breaking the vase case is not proportional or fitting. And any attempt to change his future behavior via blame in the tripping case seems inappropriate: What sort of behavior would I be trying to move him toward? How much more careful might I expect him to be? If I attempt to encourage him to search carefully for anything he might trip over every time he carries anything fragile, I would be demanding behavior that goes above and beyond the reasonable standard of care. Such behavior might be counterproductive and inefficient, all things considered. If my brother were a UPS delivery person, for example, it could slow his deliveries down to a crawl. On our limiting retributive view of punishment, which will be discussed in chapter 10, proportional expressions of blame in negligence or lapse cases may aim to bring a person's behavior up to the standard of a reasonable person's behavior.[2] However, if I blame my brother for tripping over the rug and ask him to be more careful, I am asking him to exhibit behavior above a reasonable standard of care.

The category of negligence in the criminal law echoes folk practices of holding persons responsible for lapses—so much so that we feel the burden is on others who wish to claim the two types of holding responsible for lapses do not exhibit the same structure. The Model Penal Code's[3] definition of negligence states that a person acts negligently "...when he should be aware of a substantial and unjustifiable risk" of criminal harm. Further, "[t]he risk must be of such a nature and degree that the actor's failure to perceive it, considering the nature and purpose of his conduct and the circumstances known to him, involves a gross deviation from the standard of care that a reasonable person would observe in the actor's situation."[4] The law holds a defendant responsible for harm he or she should have foreseen, even if the defendant does not actually foresee it, where the failure to foresee the harm grossly violates the standard of care a reasonable person would exhibit. In such cases, blame seems to be fitting and might further serve the aim of encouraging the person blamed to bring his or her behavior up to a reasonable person's standard in the future. For example, Bert violated a reasonable standard of care when he did not pay close enough attention to his parenting schedule. A reasonable person would have made sure to be clear on when he was expected to take custody of his children. Bert's lapse was thus like forgetting to pick up the vase and is unlike tripping over the rug.

All available evidence in Bert's case indicated he possessed an MWS (minimal working set of executive functions) and thus capacity responsibility. Bert had been a construction worker for the same firm for almost 10 years; he had had almost no contact with the criminal justice system before being arrested for child neglect; and he had no known history of mental illness. As we discussed in chapter 4, Hart argues that because capacity responsibility concerns a general ability to understand and conform one's behavior to rules, it is a foundational requirement for the efficacy of law. The criminal law requires that persons can understand the rule of law and make decisions which abide by legal rules—if they could not, the criminal law would be a useless human institution. We have argued that an MWS comprises an agent's ability to understand the demands of morality and the law, and the volitional control to behave in accordance with those demands.

In some cases, a synchronic assessment of whether a defendant had these capacities at the time criminal harm is caused is sufficient to provide the background conditions against which one may seek specific mental states regarding criminal harm caused to attribute legal liability responsibility for some action. However, where there are no synchronic mental states connecting an agent to the criminal harm he or she caused (because the agent did not know of or intend the criminal harm), it seems moral desert and responsibility rest upon possession of capacity responsibility over time, in conjunction with an assessment of what a reasonable actor would have known or intended. Bert seemed to have the capacity to understand moral and legal rules and the volitional control to behave in accordance with those demands over time. If we consider the case from a diachronic perspective, we claim that because Bert possessed an MWS it is appropriate to say he *should have known* that he was causing his children to be neglected when he went away for the weekend, and that he should have taken steps to make sure he met his parental responsibilities. Bert is responsible on our theory because he possessed an MWS (criterion 1), failed to prevent the consequence of harm to his children (criterion 2), and his executive processes should have played the appropriate role to prevent the harm (criterion 3).

One way to envision what we mean here is to think about the way in which the capacities grounded in an MWS, such as top-down attention, planning, long-term memory, inhibition, and control over the emotions, can be exercised over time to set priorities, develop a life plan, and create dispositions that constitute a person's character. The importance of diachronic self-authorship and control to ethical behavior is widely recognized in the philosophical literature. As discussed in chapter 3, executive processes, located primarily in the prefrontal cortex, may directly manipulate more inflexible brain processes via something like a process of Aristotelian habituation of traits (Aristotle 1985). Agents can use their executive functions to exercise top-down diachronic agential control over mental processes and the way in which they generate action, and we agree with Roskies that this diachronic self-authorship is important to a compatibilist account of responsibility that preserves real agency (Roskies 2012). Executive processes allow agents to perform indirect manipulations on themselves by engineering their environment to elicit certain behavior (e.g., setting a reminder, keeping a calendar) or by manipulating subconscious mental processes that generate action (e.g., training themselves to assign higher levels of emotional salience to certain environmental cues). Other examples of diachronic interventions are persons who want to lose weight avoiding having sweets in the house, persons who train themselves to slow down and count to 10 before they speak when they are angry, persons who want to exercise asking a friend to drop them off at work so they have no way home but to walk, and persons who want to make sure they pay attention to their children or partner turning their phones off when they arrive home.

Given that Bert seemed to possess an MWS, and thus these diachronic capacities for self-authorship and control during his time as a parent, he might have trained himself to pay close attention to actions that might affect his children by imagining their faces when a decision that affected them arose. He might have utilized a hard rule of writing down every aspect of the schedule with his children to avoid lapses. Knowing that he tended to tune his ex-girlfriend out to avoid conflict over the children, Bert could have practiced focusing on her communications about their children nonetheless. Bert also could have engineered his environment such that he was less likely to forget his parenting schedule (he could have set a reminder or kept a calendar) and practiced the process of deliberative self-control by slowing down and mentally reviewing his schedule before making decisions that might involve his parenting schedule, such as deciding to leave town. Such diachronic interventions would have led Bert to avoid a lapse with regard to his parenting schedule.

These capacities for diachronic self-interventions and control, we believe, ground basic desert for persons with such capacities who cause negligent criminal harm, including Bert. Because Bert had the capacity to understand and behave in accordance with the law regarding his parental obligations in the time leading up to his arrest, he is deserving of blame when he did not meet the law's demands by failing to meet a standard of care. Further, we believe Bert's guilty verdict, and his subsequent sentence of community service and parenting classes, seems proportional to his wrongdoing (fitting or just deserts). Bert's punishment may also serve the forward-looking aim of deterring him from such lapses in the future.

However, it seems that on Levy's view Bert is not responsible for harm caused by his lapse. Bert was not consciously aware of the moral implications of his decision to go to Las Vegas—he was not aware that his children would be abandoned. Therefore, Bert's neglect was not related to his "thick agency" on Levy's theory; Bert's consciousness did not have the right kind of relationship to or control over his action of abandoning his kids for Bert to be held responsible. Even if the lapse seems in keeping with what we know about the agent's attitudes—even if we know from Bert's past behavior that he doesn't care much about the well-being of his kids, especially when there is something pressing he would rather do—Levy claims we can't assume the lapse is causally related to his attitude of noncaring. Only when an action is related to a conscious attitude can we be sure the attitude is endorsed by its holder. One exception Levy might allow is responsibility for lapses where an agent shows a clear pattern of lapses with regard to a certain object (Levy 2011). So, if Bert had repeatedly forgotten to come home to his children, Levy might hold him responsible. However, we feel that even if no such clear pattern exists in Bert's case, Bert ought to be held responsible.

Levy claims that people ought not to be held responsible for "one-off" lapses. We claim, however, that the appropriate level of responsibility applied for a lapse often depends not only on whether the agent exhibits a pattern of lapses, but on whether the agent has important reasons *not* to lapse. If Sarah forgets to take out the trash, her wife may be annoyed, but she may also refrain from holding Sarah responsible, especially if there were extenuating circumstances (e.g., maybe Sarah was on her way to an important job interview). But if Sarah forgets to come to her wife's biopsy, her wife is very likely to hold her morally responsible even if Sarah has never exhibited this sort of lapse before. In other words, some one-off lapses may be excusable, but others—where there are important reasons to remember, as is likely to be true where the law demands that persons exhibit a reasonable standard of care—seem like clear cases where an agent should be held directly morally responsible for the forgetting.

Levy's focus on synchronic conscious awareness means his theory does not hold morally responsible many persons who might be found guilty of criminal negligence under current U.S. law. Levy may either claim the law is wrong to hold such persons responsible or attempt to support criminal punishment of lapsers whose behavior leads to criminal harm even if they are not morally responsible. We feel that the former is wrong not only because we think lapsers are morally responsible, but also because allowing lapsers to go free goes against important forward-looking aims of law, including regulating and influencing behavior such that persons are law-abiding. Fines and criminal punishment are important means by which the law motivates citizens to be law-abiding, and lapse cases seem ripe for the imposition of punishment as a means to encourage law-abiding behavior. This is because lapses that harm others or violate the social order can often be avoided. For example, the imposition of punishment for forgetting to care for one's children would seem to be an effective means to encourage some parents to take their parenting duties seriously. Parents like Bert who share custody ought to ensure they do not forget the dates when they have custody, and knowledge that one may be faced with a criminal penalty for forgetting would seem to be motivating to at least some potentially forgetful parents.

As we have discussed elsewhere, however, the latter option—criminal punishment of those who are not morally blameworthy—also seems problematic (Sifferd 2016b). Levy's theory essentially severs the connection between a negligent lapser and the criminal harm he or she causes: It was not Bert that caused the criminal harm of his kids being abandoned, but some state of affairs not attributable to Bert. As argued in chapter 6, Levy would claim that Bert's lapse was not due to *him* in an important sense. Bert's lapse did not issue from his agency, so the harm related to this lapse is nonculpable. According to the MPC, and all retributive and most hybrid theories of criminal punishment, the primary aim of punishment is to deliver just moral deserts. Without minimal moral culpability—if Bert did not do anything wrong—he cannot be found guilty of a crime such that he may be criminally punished.

Levy might attempt to avoid both of the difficult outcomes by claiming he can attribute responsibility to Bert via tracing conditions. Tracing allows responsibility for some conduct missing a conscious mental element (awareness of the moral implications of the act) to be traced back to a decision or action that does possess a conscious element. In Bert's case it seems we might use tracing to look for a decision or act wherein the harm he caused to his kids (forgetting his parenting schedule) was reasonably foreseeable—for example, he might have consciously decided not to write down the schedule. However, we agree with Matt King (2009), who argues that the hallmark of negligence cases is not a conscious disregard of a risk (this is the standard for recklessness, not negligence), but the failure to exercise a standard of care, which often manifests as a failure to pay attention or a failure to take note of the risk. Tracing only works, says King, when a prior decision or action meets two conditions: (1) The agent is responsible for that earlier decision or action, and (2) that earlier decision or action caused the agent to fail to satisfy the conditions of responsibility for the later action (King 2009). This is fairly easy to do in cases where an agent makes a choice that impairs his later agency, such as drinking too much or sleeping too little. In standard cases of negligence, however, King argues that it is often too difficult to trace back to a prior conscious choice that caused the subsequent lack of proper attention or care. That is, in Bert's case, even assuming there was a conscious decision not to write down the schedule (and Bert explicitly argued that this was not the case), this earlier decision was not necessarily the reason why Bert did not remember it was his weekend to take care of his kids. He might have forgotten even if he had written the schedule down.

In the end, we feel our theory, and not Levy's, can vindicate both folk intuitions that persons are responsible for serious lapses, and the criminal category of negligence.[5] This is because we do not claim consciousness of the moral implications of an act are necessary for a person to be responsible for that act. Instead, we claim responsibility rests upon the operations of executive functions (which, admittedly, are often active during periods of synchronic consciousness). Persons who possess an MWS are responsible because they can exercise diachronic agential control to be law-abiding. Capacity responsibility rests upon both diachronic and synchronic cognitive abilities—grounded in executive functions—allowing persons to understand and behave in accordance with legal and moral rules. In negligence cases, legal liability responsibility requires diachronic capacity responsibility (grounded in the operations of an MWS over time) plus violation of a legal standard of care; in most other cases, legal liability responsibility requires capacity responsibility (grounded in the operations of an MWS over time) plus a specific synchronic mental state with regard to the criminal harm, operated upon by executive functions, related to a voluntary act.

Tracking Responsibility: Dreaming, Drinking, and Flow

Even in cases where the agent had explicit states of consciousness indicating responsibility, such as planning and intending a crime, we claim it is the executive processes that are doing the heavy lifting, not consciousness. Further, we will argue that the presence of executive activity tracks our assessments of responsibility better than the presence of consciousness.

In this regard, the dreaming brain proves to be a fascinating arena. Fortunately, what happens in our dreams isn't real, and it would be difficult to argue that we are responsible for our dream actions. But it seems that we are conscious of some of our dream states. What if someone, while dreaming, performed real actions that caused real harm?

The Parks Case

In the middle of the night, on May 24, 1987, Kenneth Parks drove 14 miles through Toronto's suburban streets to his in-laws' house where he attacked them, seriously injuring his father-in-law by choking him to the point of unconsciousness, and killing his mother-in-law by stabbing her with a butcher knife he got from their kitchen, as well as beating her with a tire iron he got from the trunk of his car. According to Parks, however, he had actually been *asleep* during the drive over and even during the attack itself and only awakened during the drive home. Parks said he awakened, saw blood on his hands, and then drove to the local police station and said, "I think I have killed some people … my hands" (Broughton et al. 1994). It appears that Parks had a special type of sleep disorder, REM behavior disorder (REMBD), that can cause people to get up out of bed and perform actions while technically still asleep.[6] Parks showed abnormal EEG in the sleep lab, as well as periods of what sleep researchers call "partial awakening." Parks was acquitted by a jury. His sleepwalking claim, bolstered by the absence of a motive, as well as his truthfulness and consistency throughout the trial, convinced the jury he was not responsible.

REMBD can be distinguished from normal sleepwalking since the latter happens the vast majority of the time during non-REM sleep. People with REMBD are said to act out their dreams—which are themselves typically described as vivid, violent, and frightening—because the normal mechanism that prevents dream states from actually causing behavior is malfunctioning. This results in "overt motor behaviours that usually represent the enactment of dreams" (Matar and Lewis 2017, 262), such that, when awakened, people are able to recall the dream they were having, which will match the actions they were performing. It looks as if what is happening is that the visual system is functional at some level, and the brain is able to produce conscious states of a special type, mixing together perceptual content with dreamed content. REMBDers are somehow dreaming what they are seeing. One woman dreamed she was swimming in a deep river, getting tired, and in danger of drowning. This was in fact the case, and luckily she was able to get to shore (Brogaard and Marlow 2012).

We agree with Levy that Parks is not responsible for the harms he caused because he fails to meet the mental conditions for responsibility. We disagree with Levy, however, as to the exact cause of that failure. Levy argues that Parks is not responsible because he lacked states of conscious awareness of the moral significance of what he was doing (Levy 2014, 89). Levy's primary reason for this is that Parks had no conscious awareness of *anything* during the attacks, presumably because he had no states of access consciousness the whole time. Against this, we will argue that Parks is not responsible for different reasons. Parks fails to meet two of our criteria for responsibility.[7] He did not possess a minimal working set of executive processes at the time of the crime (criterion 1), and Parks's actions do not have the executive engagement in their causal history needed to make him responsible for the murder and assault (criterion 3).[8] Parks fails both of these criteria because the executive processes are substantially shut down during sleep, as we will see. Because most executive processes are inactive during dreams, Parks's actions cannot be considered planned or intentional, despite perhaps appearing that way to an outside observer.[9] As he stands there in his in-laws' bedroom wielding a butcher knife, Kenneth Parks is not a competent moral agent.[10]

Several imaging studies of dreaming people have confirmed the finding that the prefrontal lobes are substantially shut down during sleep (Muzur, Pace-Schott, and Hobson 2002, Dang-Vu et al. 2010).[11] Some imaging studies of brains in the midst of REM sleep appear to show significant decreases in activity in the dorsolateral prefrontal and parietal portions of the cognitive control network while continuing to show activity in relevant portions of the anterior cingulate (e.g., Desseilles et al. 2011). We are indeed remarkably unthoughtful in dreams. We don't problem solve, except in bizarre or primitive ways. We not only don't work to resolve conflicts, but we are untroubled by them. The dreamer, and presumably the person in an REMBD episode, cannot plan, cannot monitor, cannot inhibit, and cannot resolve inconsistencies. Dreaming people also don't inhibit themselves from doing illegal or unethical things very well. Things mainly just tend to happen, including "our" own actions. We certainly tend to get strong emotional reactions to dream events, and the imaging studies show plenty of activity in the brain's emotional systems (see Maquet 2000, Maquet et al.

2005), but again, there is no executive management of these emotions and their effects on us and what "we" do in the dream. Therefore, Parks's drive to his in-laws' home must have involved something similar to the driving on autopilot we all sometimes do. There is also no real evidence of planning in what Parks did when he got to his in-laws'. In general, the fact that the cognitive control network is shut down means that a person does not have control over what happens in his or her consciousness, and over how those conscious states lead to actions. None of the actions such a person might commit will have executive "approval" or even executive engagement.

Given that two of the three core areas that make up the cognitive control network are deactivated during REM sleep, it seems unlikely that executive activity could occur—dreamers might meaningfully be said to suffer from dysexecutive syndrome. The set of executive processes that are active during dreams, if any, is not the same as the set of executive processes active during waking novel cognition, and it does not constitute a minimal working set. In terms of the self, the dream self is not the same as the normal waking self. This is easy to observe if you compare what "you" do and how "you" react in dreams with what you do and how you react in waking life. You will find many inconsistencies. Your dream self does things you would never do. It reacts in ways that you would never react. It can behave irrationally. It is completely unsurprised when dead people or famous people appear in your dreams, or your car turns into a boat, or your office mate turns into a TV character. Your dream self has beliefs that you do not have. Most of the time, this dream self seems to lack any useful access to your memories and to your normal waking abilities. For example, you know how to run, but in your (bad) dreams you pump your legs furiously but get nowhere, or you slip on ice that magically appears. And instead of being dumbfounded by this, and struggling to explain it, without executive processing you just keep trying to run.

Can Levy Exculpate Parks? It is not entirely clear that Levy's theory is able to exculpate Parks. Levy requires consciousness of the moral implications of an act. According to Levy, this involves information being personally available to Parks and poised to be reported. Could Parks have made a report on his actions while asleep? The answer is not clear. Certainly, people in REMBD episodes will talk, and Levy notes that Parks was "a chronic sleep talker and occasional sleepwalker" (Levy 2014, 71). Suppose he is able to give coherent (but perhaps false) answers to questions about what he is doing and why without awakening. We do not see that this carries much weight in making him responsible. As for being conscious of moral implications, also required by Levy's theory of responsibility, dreams are a type of conscious state that can involve awareness of the moral implications of something one is actually doing. For example, if during the killing, in his dream Parks believed he was stabbing his mother-in-law because she was threatening the lives of his family, he apparently *does* have a conscious state of awareness of the moral implications of an action, that is, the actual (and dreamed) action of stabbing her. His belief is false, but recall that Levy states that this awareness does not have to be factive (Levy 2014, 37).

The Nature of Dream Consciousness Exactly what was happening in Parks's mind that night is not known. We do know that people in the midst of an REMBD episode in general are dreaming but that they are also able to make use of some sensory input and are sometimes able to perform nonroutine actions. Levy's fundamental position appears to be that there is no consciousness present in Parks during the attacks. He categorizes Parks's behavior that night among other behaviors which are "driven by nonconscious states" (Levy 2014, 78). In addition, though, perhaps because of the unclarity about what was happening in Parks, Levy also allows that people in REMBD episodes may have a sort of consciousness. He says that "during sleep violence they are conscious, but of self-generated contents" (Levy 2014, 78n). Such people, says Levy, "may utilize perceptual information, but only in certain ways, and only information that falls within a certain range" (Levy 2014, 78). This again seems to offer the possibility that Parks might have conscious states. In our analysis, we will address the range of possibilities with regard to Parks's consciousness, from the claim that Parks had no consciousness at all, to the idea that he had conscious states, but still not the right conscious states for responsibility.

The things Levy says about dreaming consciousness and autopilot driving work to show that the type of consciousness he says matters to responsibility—access consciousness—does not line up well with some of our core intuitions about what consciousness is. Those who believe that all consciousness is access consciousness have trouble categorizing dream states and autopilot states as conscious states. The dreamer does not seem ready to report the contents of his or her states, or use the information to problem solve, as would be the case with a normal state of access consciousness. But surely dream states *are* a type of conscious state.[12] They contain colors, shapes, sounds, feels, and emotions; there is something it is like to experience them. The brain's emotional systems tend to show high rates of activity during dreams (see Maquet 2000, Maquet et al. 2005), and we can all attest that dream experience can contain powerful emotions, sometimes even reactive emotions to actions we see as wrong.

Levy takes a similar line in the case of mindless driving. We have all had the experience of driving "on autopilot": We complete a very familiar drive with our minds on something else. We arrive home and realize that we have not really been consciously aware of the drive. Levy says that when we are driving on autopilot, we "act in ways driven by action scripts, in the absence of very much consciousness of what we are doing" (Levy 2014, 76). He is perhaps correct that the visual states of the traffic are not states of access consciousness because the executive processes are inactive or working on something else. They are not accessing or monitoring the person's (phenomenally) conscious visual states, which is why it can take a moment to arouse them. Normally, when executive processes are monitoring—for instance, when you are looking for a certain thing, such as waiting for the light to turn green, or listening for a certain sound—you realize instantly when you see that thing or hear that sound. However, there are (phenomenally) conscious visual states in your brain when you drive on autopilot, something that it is easy to verify from your personal experience. The other cars have colors and shapes; these colors and shapes are bound together with sounds, the sound of your engine or the radio, even when your primary conscious focus is on an internal dialogue having nothing to do with the traffic. If there were really no conscious states when driving on autopilot, there would be no colors, no sounds, and this is not what happens in autopilot mode. It looks as if the concept of phenomenal consciousness not only matches the neuroscientific data better than the concept of access consciousness, but in the cases of dreaming and automatic driving, it better matches our folk intuitions.

It seems to us that dream states can sometimes be (conscious) states that contain representations of facts along with representations of moral valence (e.g., I am stabbing my mother-in-law; this is good because she is the devil). If this is right, then there is a clear way that Parks could satisfy Levy's consciousness criterion. Protecting loved ones from danger or harm is a recurring theme in the dreams of people with REMBD (Schenk and Mahowald 2002). Walker and McCabe (1968) document a case in which a man killed his infant son during an REMBD episode, thinking he was smashing a "white beast" that had come up through the floor. Contrary to his stated position, Levy might be stuck with claiming that Parks *is* responsible.

Levy correctly notes that typical sleepwalking is done in non-REM sleep, but that sleep violence is correlated with dreaming. "This evidence suggests that during somnambulism, the sleeper is not conscious of anything, whereas during sleep violence they are conscious, but of self-generated contents, rather than of the external environment" (Levy 2014, 77–8n).[13] If the person in the REMBD episode committing the violence can be conscious of things, then Parks could be conscious of the moral significance of his assaultive actions. It is also important to note that the conscious states the person is experiencing are not just any random self-generated dream contents. They must be visual states of a sort, of the person's current surroundings. Otherwise, how did Parks drive and kill successfully? Indeed, REMBD behaviors frequently require perception as well as some skill. As we noted above, it may be that people in this state have a type of conscious vision. This could be what allows them to perform novel actions while their frontal lobes are asleep. Both our theory and Levy's theory claim that something more than routine action is needed for responsibility. Levy says that the sort of behavior Parks exhibited is behavior generated by "action scripts." This may be right, but people following such scripts are not behaving blindly. The scripts are designed to be interrupted, based on what the person is consciously perceiving.[14]

Levy notes that, if there is perception during an REMBD episode, it is within a restricted range and can only lead to a restricted range of behavior. But this restriction, we would say, has nothing to do with consciousness. It is due to the absence of the executive processes and their ability to deal flexibly with novel situations. Presumably, these executive areas are shut down in Parks when he is committing the crime, but this does not mean that there are no conscious states in Parks's brain. There may be conscious visual states, for instance, in his brain when he is attacking his in-laws. These states can be assigned the content that he is stabbing someone, that she is begging him to stop, even that what he is doing is right, and so on. However, these states are not engaged in causal interaction with the cognitive control network, and this is why we balk at saying that Parks is conscious that he is stabbing someone.

"Parks Is Not Conscious that He Is Stabbing an Innocent Person" According to our view, it is true that Parks is not conscious that he is attacking his in-laws, but not because of a failure of consciousness to be present. Rather, it is *Parks* who is not fully present. This is because the executive processes are not available. Levy likewise says, "He is not conscious *that he is stabbing an innocent person*; he is not conscious *that she is begging him to stop*, and so on" (Levy 2014, 89). This is correct, but not because the man doing the stabbing has no conscious states, or even no conscious states about the moral significance of his actions. On our theory, since the executive processes are inactive in Parks's brain, and they cannot be engaged in the appropriate way, he does not have a minimal working set of them at the time of the crime. We are sensitive to this when we assess statements describing what Parks did that night. What makes the sentence "Parks is conscious that he is stabbing an innocent person" false is a tacit reference failure of the sort we described in chapter 6. There is a failure of the noun phrase to refer to both the external referents and the mental referents. At the mental level, we are not tacitly referring to Parks's normal, waking cognitive control network, as we should be in making an ascription like that.

Levy agrees that Parks is not conscious of his actions but attributes this to a failure of certain conscious states to be present. Parks fails to be aware of the moral significance of his acts, according to Levy, because he is either not conscious at all or not capable of the right sort of conscious states. In analyzing the claim that Parks is aware of x, all of Levy's focus has gone into "aware" and "x" but none into the "Parks" portion. He is right to say that during the REMBD episode the sleeper, considered as the normal waking person, "is not conscious of anything," but this is because the executive processes are shut down, not because there is no consciousness. The sleeper's actions are caused by conscious states; they just aren't fully *his* conscious states. One way to see this is simply to remove the name Parks from the subject position of our claims and determine what happens to the truth of the claim (or its correctness and completeness). There are then two different claims: first, that Parks is conscious of attacking his in-laws, and, second, that there are conscious states in Parks's brain that represent him attacking his in-laws. The first claim tacitly refers to Parks's executive processes. The second does not. The first claim is false, and the second is true, according to our hypothesis.

Because he does not appeal to something like the executive processes, Levy has no obvious way to differentiate the conscious dream states of people in REMBD episodes from normal waking conscious states. The problem with this is that the difference is crucial to responsibility. The main difference between the two, we submit, is that the executive processes are "asleep" or inactive during the REM behavior episodes, but normally active, or able to quickly activate, during waking states. The dreamer's conscious states are not interacting with his or her executive processes. At the very least, Levy has no way to differentiate the two types of conscious states that can be put into neuroscientific terms. He might say that the dream states are not personally available because the person is asleep. But we do not know what the person is, in his account, other than just the full organism.[15] Recall that Levy accepts the global neuronal workspace theory of consciousness, but that "workspace" was used ambiguously to refer either to just the workspace buffer or to the entire workspace system, including the executive processes. This appeared to cause Levy to unknowingly refer to executive areas and activities in certain of his claims about consciousness. As an example of getting the benefit of referring to the executive processes by using a noun phrase, like "the agent," Levy says, "Only when the agent is conscious do a very broad range of internally and externally generated cues modulate behavior" (Levy 2014, 79).

Levy says that Parks is not conscious of the moral implications of what he is doing. We say that Parks, the normal awake person, is not conscious of anything during the attack. The problem is not a lack of consciousness but a lack of executive processing. *Parks* is not conscious that he is stabbing his mother-in-law. If I am not conscious that I am actually *doing* x, I am unlikely to be conscious of the moral implications of doing x. There are conscious states in Parks's brain that produced the murderous actions, but they are not properly *his* conscious states and they have no possibility of receiving executive approval as long as the episode continues. Thus, we cannot say that *Parks* is aware that he is stabbing and killing his mother-in-law, even if there may be a conscious state in him that represents these actions. In addition, there is no reason to expect that Parks could exercise his capacities to understand moral and legal rules and to inhibit behavior so as to be law-abiding while in the midst of an REMBD episode.

Alcoholic Blackout

Compare Parks to someone who kills during a state of alcoholic blackout. In neither case is the person himself aware of what he is doing at the time. Another way to say this is to note that their executive processes are substantially disabled at that time—they do not possess a minimal working set of them. There are some similarities between the two conditions: People perform actions that are not routine and can be fairly complicated with limited memory of the events after they wake up. There is a crucial difference between the two, however, one that bears on their responsibility for what they do. The person in a state of alcoholic blackout knowingly and voluntarily put himself in that state, in virtually every case. By contrast, the person with REMBD has no reasonable way to prevent his episodes. He is not, in our judgment, more culpable if he has done unethical things during previous episodes, because he must sleep—although he ought to take reasonable precautions if possible. The alcoholic does not have to drink, though, and he is even more culpable if he knows he has done bad things in the past after drinking too much and blacking out. If it is the first time it has ever happened to him, his culpability is less, but he still bears significant culpability. There is an executive failure to inhibit the continued drinking. This inhibition could take a variety of forms, from a simple case of prudently limiting alcohol consumption to an amount well short of intoxication in the moment to a more diachronic case in which the person gradually trains himself over time to limit his intake of alcohol.

The Unwilling Addict

People with REMBD can commit actions that have not received, and cannot receive, executive approval. Some drug addicts also seem to engage in behaviors of which they themselves do not actually approve. Once we know that Pete explicitly disavows any desire for heroin, and tells us of his unending efforts to quit, and so on, we may judge that it is wrong to say, "Pete wants heroin." However, one can say, "There is a craving for heroin in Pete," and when the craving makes him go buy heroin, we say the craving was an overwhelming one. Pete's cravings for heroin not only lack "executive approval," but they have elicited vast executive disapproval. The cravings are not really *his* cravings. Inhibitions have been repeatedly directed at them, with varying degrees of success.[16]

Similarly, the person suffering from alien hand (or anarchic hand) syndrome emits behaviors that look like intentional actions (Scepkowski and Cronin-Golomb 2003). One patient's hand tried to hit and strangle her, while another patient's hand reached out and grabbed at his fellow bus riders. However, these should not be counted as these people's actions because the actions never had executive approval or even exposure to the executive processes before they occurred. The sufferer of this odd and fortunately rare syndrome will say, "*I* didn't do that; my hand did." "Actions" undertaken without executive approval are not *our* actions.

Levy is perhaps alluding to this phenomenon when he says, "It is for this reason that we can find ourselves 'wanting' addictive drugs, despite neither enjoying their effects nor taking their consumption to be in any way valuable" (Levy 2014, 98). Any "wanting" for heroin has not received executive approval. Granted, it certainly seems like there is something in Pete, a desire, a craving, some sort of wanting, that is moving him toward procuring heroin. Anyone who didn't know him and observed his behavior would conclude that he wanted heroin. A better way to describe the situation is to say that there is a sort of wanting in Pete, but that *he* does not want heroin. There are wantings in us that are not *our* wantings. Pete has not approved of this wanting; indeed, he strongly disapproves of it. Similarly, the right way to describe implicit attitudes is as mental states (sharing some features of attitudes) that are not *ours* because they have not passed executive review.

Bert, the neglectful father we discussed above, is like the unwilling addict in that his executive functions are available but fail to act. Bert's executive processes failed to activate and initiate behavior that would have prevented the neglect. While the executive processes can be activated by all manner of thoughts, perceptions, or even memories, there appear to be a large set of cases in which the executive processes are activated by the autonomic system. Some of these cases could be described as activation by emotions. We feel that Bert should have worried more about his parenting schedule; he should have been cued by his emotions to pay closer attention. Conversely, failure of the executive processes to activate is typically traceable to the person's attributing a low value to the relevant situation. Thus, we take instances of failure to activate to be evidence of uncaring—for example, "You must not love me very much to forget my birthday." Bert wasn't sufficiently alarmed at the prospect of his kids spending the weekend alone, and we might think the unwilling addict is not sufficiently alarmed by his addiction. In such cases, we often try to correct people's failure to use their executive processes when they should by supplying them with a rousing motive, such as a threat to punish.

However, as we noted in chapter 3, it's not just that the person's executive processes failed to activate, but it's that they failed to activate when they *should* have (from our criterion 3). The power of the executive system to inhibit and shape behavior has distinct limits, especially in cases like addiction. In cases of addiction, the first and possibly second time a person takes a drug, we may say the person's executive functions should have engaged to inhibit the action of taking the drug. However, as the strength of cravings grows, the executive may become less and less sensitive to other reasons to act in certain situations, and the person may become unable to inhibit certain actions. In some cases, the executive can be trained and strengthened to better manage the cravings; in other cases, medicine can decrease cravings to give the executive better control over behavior. In general, with cases of addiction the executive processes are on the front line of the unwilling addict's battle for control over his or her behavior, and any adjunctive aids, such as therapy or medication, will be vastly more effective if done together with concerted executive activity.

Flow Experiences

Our ability to observe the working brain is now powerful enough that the question of whether flow experiences involve executive processing can be approached with straightforward empirical techniques. We suspect that in one type of flow experience what happens is that the executive processes initially establish the perception-action channels through which the flow is going to pass. As we noted in chapter 3, the executive processes must supervise the diachronic establishment of habits by planning, attending, task switching, and so on. The expert point guard in basketball (the team's primary ball handler and playmaker) spends a lot of time setting up the sort of flow that happens so quickly and naturally during a game as this player perceives the positions and movement of both teammates and the opposition. The point guard had to practice several moves, dribbles, pivots, and passes, so that he or she could then put them together in real time, improvising depending on what the defense does. The vast majority of the time, the point guard is not engaged in executive-level decision-making during a game. This player is simply allowing his or her perceptions to guide his or her actions in real time, but in ways that can't really be called routine.

Similarly, jazz musicians practice to improvise. When they do improvise in performance, it is often not the first time they have played that series of notes, or arpeggio, or chord. But it might be the first time they have used that series of notes in that sort of musical context, as created by the other musicians, as well as other factors. People are still responsible for actions committed during flow states, on our account, because those actions (or their constituents) were knowingly practiced and perfected—their MWS was crucial to their establishing a "flow"—and, unlike the sleepwalker, persons in a flow state can engage their executive functions if this becomes necessary.

Conclusion

We believe persons are responsible for lapses that cause harm where they possess an MWS in the time period prior to the lapse. A person with a minimal working set of executive functions has the capacity to manipulate his or her agency and dispositions to act over time such that he or she is more likely to realize desired behavior and avoid undesired behavior. This capacity grounds responsibility in lapse cases.

Typical sleepwalkers are not dreaming for the most part, but people in the midst of an REMBD episode are. If we regard dreams as conscious states,[17] someone with REMBD such as Kenneth Parks actually could have a conscious state that was about the moral significance of what he was doing (an action he may have been performing both in his dream and in reality). That means a consciousness-based theory such as Levy's might have to find Parks responsible. Our position is that Parks would be right if he were to claim: "*I* didn't intend to kill her, that wasn't what *I* wanted. *I* was asleep!" Parks was not in possession of his minimal working set of executive processes at the time of the attacks. Whatever else happened in Parks's brain that night, his actions did not pass executive review, or get "executive approval," because his set of executive functions was not operating.

8 Responsibility under Development

How Did We Get Here? The Puzzle of Juvenile Responsibility

We have offered an account of responsibility according to which, if an agent is responsible for some action, it is necessary that the agent possessed an MWS (minimal working set of executive functions) at the time of the action. For many persons facing the prospect of legal punishment, there is no serious question about whether they satisfy this condition. Such persons are clearly responsible agents (or were at the time of their alleged offense); the open question is whether they are responsible *for some particular act*. We have laid out our answer to that question in chapters 3 and 4.

However, as the criminal law already recognizes, there are offenders to whom responsibility may not be properly ascribed at all, insofar as they fail to satisfy what H.L.A. Hart (1968) calls "capacity-responsibility" (discussed in chapter 4). In this chapter and the next, we will argue that, just as the folk-inflected concepts of moral and legal responsibility implicitly refer to executive functions, certain legal excuses such as juvenile status and legal insanity aim implicitly at offenders who fail, for various reasons, to satisfy the MWS condition. We begin with the puzzle of juvenile responsibility—which, even among the tangle of difficult philosophical issues pertaining to responsible agency and action, seems particularly tricky. At its borders, the vast territory covered by the term "juvenile" seems easy enough to navigate: If any human being is exempt from blame because of a lack of capacity responsibility, a newborn infant is so exempt. Skeptics aside, it seems that if anyone satisfies capacity responsibility, a neurotypical adult does. But when and how, in the span of years stretching from birth to adulthood, do the capacities for responsibility emerge?[1]

Furthermore, considering the responsibility of juveniles yields intuitions that exist in tension with one another. We often think of young people—especially the preadolescent—as innocents, unable either to control their actions or grasp the consequences of those actions. Their characters and dispositions are still inchoate, and the genetic and environmental factors shaping those dispositions are not up to them. On the other hand, children appear capable of *some* degree of control; they are able to delay gratification in favor of a delayed reward, as Mischel's famous experiments showed;[2] they evince emotions of guilt, remorse, and shame from a very early age, and these emotions may reflect a genuine, if still maturing, commitment to morality.[3] These early signs of a moral sense need to be nurtured. Influenced by Stephen Darwall (2006), Krista Thomason (2016) argues that too much reluctance to hold children responsible for their actions risks failing to honor that commitment, thereby damaging children's nascent connection to the community defined by shared norms.

Recall chapter 1's description of the case of Dominic Ongwen, kidnapped around the age of ten from his Ugandan home and forcibly conscripted into the Lord's Resistance Army, Joseph Kony's guerilla force. Permanently cut off from his family and former life, Ongwen suffered continual abuse and brutal indoctrination; through luck and shrewd ruthlessness, he survived and climbed to a position of authority in the LRA. By his early 20s he had witnessed and participated in an unknown number of atrocities, and as a senior commander in the LRA, he is accused of planning military operations marked by pillage, rape, murder, and sexual slavery. He even kidnapped and forcibly conscripted child soldiers of his own, allegedly, training them to kill as he had been trained.

What can we blame Dominic Ongwen for? He seems to us a victim and a monster all at once—or, perhaps, a victim who became a monster. Many of his actions and crimes were committed as an adult and were the product of deliberate planning and careful execution; the atrocities carried out under his command cry out for justice. Yet once we learn his history, those actions seem undeniably rooted in the horrific, morally distorting trauma he suffered as a boy. As of this writing, Ongwen is standing trial at the International Criminal Court. In her opening statement, Fatou Bensouda, the lead prosecutor at the ICC, eloquently captures the tensions inherent in his case:[4]

> People following the case against Dominic Ongwen ... will feel horror and revulsion at what he did, but they will also feel sympathy.... And the phenomenon of

the perpetrator-victim is not restricted to international courts: it is a familiar one in all criminal jurisdictions. Fatherless children in bleak inner cities face brutal and involuntary initiation ordeals into gang life, before themselves taking on a criminal lifestyle. Child abusers consistently reveal that they have been abused themselves as children. But having suffered victimization in the past is not a justification, nor an excuse to victimise others. Each human being must be considered to be endowed with moral responsibility for their actions.

Ongwen's case is far from typical, fortunately, but such cases drive home the urgency and difficulty of why, when, and how to hold juveniles responsible. Despite the special challenges this topic presents, the development of responsible agency has drawn the interest of legal philosophers, criminologists, brain scientists, and others. And for those interested in the nature of human agency and responsibility itself, there is philosophical value in accurately characterizing such an elusive phenomenon. Our project here, in building an account of responsibility intended to be scientifically accurate, philosophically defensible, and reasonably complete, is motivated by both sorts of purpose.

In this chapter, we will extend our executive account of responsible human agency to address the development of capacity responsibility in the maturing brain. We will first survey the changing state of juvenile justice as it tentatively absorbs and accommodates the findings of contemporary neuropsychology. Next, we will turn to those findings themselves, which, taken together, give an increasingly clear picture of how the executive functions and their neural grounds develop across childhood, adolescence, and early adulthood. With these foundations laid, we will sketch a scalar account of responsibility and recommend, as a direction for legal reforms, a graded structure of criminal responsibility for juveniles. We will close the chapter with some brief reflections on responsibility and the aging brain. The cognitive decline associated with later life in some respects mirrors the cognitive and volitional trajectory of adolescence, and our remarks on the responsibility of juveniles have implications that touch the end of the life span as well as its beginning.

Juvenile Justice in Flux

The truism that young people are naturally prone to rash and risky behavior has been around, one imagines, for as long as young people have. Over 2,000 years ago, Aristotle observed that the young "have strong passions, and tend to gratify them indiscriminately.... They are changeable and fickle in their desires, which are violent while they last, but quickly over: their impulses are keen but not deep-rooted...." (Aristotle 1984, 2213).

Even if Aristotle is right it does not necessarily imply that the young are any less deserving of blame for what they do. As Stephen Morse argues, having a character that makes bad behavior harder to avoid is rarely considered a good basis for excuse, because "[t]he law assumes that people who are characterologically thoughtless, careless, pugnacious, excitable, cowardly, cruel, and the like have sufficient general normative capacity to be held accountable if they violate the law" (Morse 2000). If the law is prepared to assign responsibility to adults who are passionate, fickle, or impulsive, then consistency would seem to demand the same assignments for young people with these character traits—unless we believed that juveniles were not *merely* passionate, fickle, or impulsive as a matter of character but were somehow diminished in the "general normative capacity" required for responsibility. And modern legal jurisdictions have increasingly affirmed this very belief, drawing a link between juveniles' underdeveloped capacity for rational self-governance and their reduced culpability (Brink 2004).

The notion of juvenile justice itself is still relatively new; separate court systems for juveniles did not emerge until the turn of the twentieth century. Before that, common-law traditions regularly assigned full, adult-level culpability to children as young as eight or nine. Now, over 100 years later, practices and doctrines governing the treatment of juvenile offenders vary widely from one jurisdiction to another. Within the United States, states differ in their definitions of the minimum age of criminal responsibility. Although the MPC gives juvenile courts exclusive jurisdiction for defendants under 16, many states instead use a cutoff of 18 (Cipriani 2009, 221–2). Juvenile offenders are generally tried in a separate system from adults, one marked by more lenient penalties and a comparatively greater focus on restorative and rehabilitative measures (Cipriani 2009). But the practice of juvenile transfer, the specifics of which differ from state to state, allows for juveniles to be tried as adults in particular kinds of cases—for instance, if the crime in question is especially serious or violent. Some juvenile transfer policies rest on purely statutory grounds (e.g., some automatic transfers are required due to severity of the crime), but often these decisions are left to the discretion of judges or prosecutors (Addie et al. 2011).[5]

The juvenile justice system emerged from movements aimed at social reform, and it has been the continual focus of reformers' efforts ever since. In the last 15 to 20 years, those efforts have been significantly shaped by the findings of neuroscience and developmental psychology, fields increasingly cited by lawyers and judges in criminal law (Farahany 2015). That same time period has witnessed a rapid growth in our understanding of how the capacities for self-control and moral cognition develop in the brain. From these two trends, a third has emerged: The U.S. criminal law has moved toward ensuring that assessments of culpability and punishment for juveniles are proportional to their differential status. Partly as a result of a series of high-profile decisions, courts have explicitly affirmed that juveniles are less deserving of criminal punishment than adult offenders, especially our harshest punishments, and that justice demands they be punished in proportion to this degree of desert (Pillsbury 2013).

The Supreme Court's momentous decision in Roper v. Simmons (2005) was a key moment in the establishment of this pattern. By a 5–4 majority, the Court found that the Eighth Amendment's ban on "cruel and unusual punishment" prohibits the death penalty for juvenile offenders.[6] Five years later the Court extended this reasoning, in Graham v. Florida (2010), to cover sentences of life in prison without the possibility of parole (LWOP) for nonhomicide juvenile offenders, and extended it further in Miller v. Alabama (2012) and Jackson v. Hobbs (2012), to ban mandatory LWOP sentences for juveniles convicted of first-degree murder.[7]

That juveniles are undeserving—or less deserving, at least—of the harshest forms of criminal punishment, such as death or life in prison, is one expression of the "jurisprudence of proportionality" emerging from these decisions (Pillsbury 2013, 887). The Court's view that juveniles are less culpable and more corrigible than adults reflects the common wisdom that "any parent knows," as Justice Anthony Kennedy wrote in *Roper*. But it also draws on scientific evidence—including data from developmental psychology and neuroscience—of juveniles' immature cognitive and volitional capacities. In Justice Elena Kagan's *Miller* opinion, she cites *Roper* in summarizing the three main conclusions suggested by this evidence:

> First, children have a "'lack of maturity and an underdeveloped sense of responsibility,'" leading to recklessness, impulsivity, and heedless risk-taking. *Roper*, 543 U.S., at 569. Second, children "are more vulnerable … to negative influences and outside pressures," including from their family and peers; they have limited "contro[l] over

their own environment" and lack the ability to extricate themselves from horrific, crime-producing settings. *Ibid.* And third, a child's character is not as "well formed" as an adult's; his traits are "less fixed" and his actions less likely to be "evidence of irretrievabl[e] deprav[ity]." *Id.*, at 570

Despite this basic rationale prevailing in the high-profile cases mentioned above, the Court's decisions hardly reflect a settled consensus on the issue of juvenile culpability. Indeed, each of these cases was decided by a bare 5–4 majority; the winner-take-all nature of such decisions should not blind us to the divisions still clearly evident in the Court's thinking. Furthermore, even *Miller* bans only the *mandatory* imposition of LWOP sentences for juveniles convicted of first-degree murder. Jurisdictions may still impose LWOP without running afoul of *Miller*; the decision simply grants lower courts the freedom to hand down a lesser sentence.

Finally, consider Justice Kagan's own remarks in the *Miller* opinion. After reviewing some of the facts surrounding Evan Miller's case—when Miller murdered his neighbor, Cole Cannon, by beating him with a bat and setting fire to his trailer, Miller was 14 years old; he was under the influence of drugs and alcohol at the time and was a regular user of both; his childhood had involved prolonged patterns of abuse and neglect, and he had drifted in and out of foster care; he had attempted suicide at six years old, and three more times afterward—Kagan argues that such circumstances ought to figure in a court's decision about what sort of punishment Miller deserves. Still, she claims, "No one can doubt" that Miller's crime was "vicious" and finds it "beyond question" that Miller "deserved severe punishment" (16). These remarks, and the ongoing controversy over these issues in the wake of *Roper*, reflect a conflict between two powerful forces: on one side, the deep social need to affirm our shared norms by responding to brutal, dangerous actions with harsh condemnation; on the other, the growing sense that at least some forms of condemnation and punishment are simply unjust or inappropriate when applied to young people, especially those whose lives have been profoundly shaped by abuse, violence, drug use, and mental illness.

To resolve this conflict, we need what has been so far lacking: a scientifically sound and philosophically correct account of how the capacities necessary for responsible action develop during childhood and adolescence. We will argue that understanding how executive functions mature in the juvenile brain enables us to explain when and why juveniles should be considered candidates for diminished responsibility and, in some cases, excused altogether from it. If the thesis advanced in this book is accurate, then the problem of how to hold juveniles responsible for their actions hinges on the extent to which juveniles can be said to have an MWS.

Executive Functions under Development

Surveying the contemporary research on executive development in "the juvenile brain" carries an obvious risk of oversimplification. The term "juvenile" covers an extremely broad age range, and juveniles are a wildly heterogeneous class. An average ten-year-old's brain and level of executive control are very different from those of an average 18-year-old. Moreover, individual juveniles develop along different trajectories, shaped by distinct genetic, epigenetic, and environmental factors; an arbitrary pair of 15-year-olds might evince quite disparate levels of ability. Absent a severe disorder, illness, or abnormality, children and adolescents clearly possess some ability to inhibit behaviors, shift their attention, manipulate items in working memory, switch between tasks, detect and correct errors, and more. However, it is equally clear that childhood and adolescence involve a gradual maturation of these abilities, and that this developmental trajectory resists being carved into easily discriminable stages. The process of executive maturation is messy, variable, and only starting to yield to scientific investigation, but developmental neuropsychology has established enough for us to accurately sketch the outlines of that process.

Aside from space constraints, there are good reasons to give comparatively more or less attention to certain roughly defined age ranges. For example, there seems to be no need to explore in detail gradations of culpability in preadolescent children. We are already strongly disposed to think of very young children as significantly less responsible for what they do, if responsible at all. And young children simply do not have frequent opportunity to cause harms or commit moral violations serious enough to warrant the forms of blame and punishment that mark the criminal law. With that in mind, we will spend relatively little time on the earliest years of life and comparatively more time on adolescence. Adolescence is the time when young people are most disposed to risky, criminal, or antisocial behaviors, which can be decidedly nontrivial in their nature and effects (Steinberg

2004). Adolescents are "more likely than both children and adults to abuse alcohol, use illicit substances, have unprotected sex, commit antisocial acts, drive recklessly, and drive while intoxicated" (Pharo et al. 2011, 970). From both practical and philosophical angles, then, adolescence is the developmental period most urgently in need of illumination, and therefore it warrants the bulk of our present attention.[8]

Infancy and Preschool (Birth to 5 Years Old)

Research shows that individual executive functions appear well before the age of three, developing along distinct but interdependent trajectories (Best, Miller, and Jones 2009, Garon, Bryson, and Smith 2008). Working memory and the capacity to manage it develop first—initially the ability to hold simple representations in mind (six months), and later to update and manipulate them (15 months). Between six months and one year, infants begin to show simple forms of the ability to delay or inhibit responses, and the response-inhibition and working-memory components begin to coordinate at around two years old. That coordination means greater ability to hold a rule in mind and use it to inhibit a prepotent response, as shown by tasks such as the "shape Stroop" test, in which children are shown pictures of small fruit embedded in larger fruit and must point, when prompted, to the small rather than the large fruit (Garon, Bryson, and Smith 2008). Cognitive flexibility, sometimes called "set shifting," is the last of the "big three" executive components to develop. Experts agree that it builds upon the foundation laid by inhibitory and working-memory capacities: One form of set shifting is the ability to change one's spatial perspective, which requires inhibiting one's own visual perspective and "loading" a different perspective into working memory (Diamond 2013, 149, Garon, Bryson, and Smith 2008).

These changes, according to many developmental psychologists, are undergirded by the continuous development of the broader attention network, especially the "orienting" subsystem (Posner and Fan 2008, Rothbart and Posner 2001, Colombo 2001), which shows significant development between birth and one year old, and the anterior attention subsystem, which "selects and enhances processing according to internal representations in part by inhibiting and facilitating the orienting subsystem" (Garon, Bryson, and Smith 2008, 35; see also Ruff and Rothbart 2001). The ability to selectively attend is a prerequisite for any executive function task, and it is "the maturation of attentional capacity" that "forms a foundation for development of EF [executive functions] abilities during the preschool period" (Garon, Bryson, and Smith 2008, 35). Indeed, there is considerable agreement that the development of attentional capacities is essential to the construction of a unified central executive as such (Baddeley 2002, Conway and Engle 1994, Kane and Engle 2003).

School-Aged Preadolescence (6 to 12 Years Old)

In the preadolescent period of childhood, specific executive functions continue to mature along distinct performance trajectories, which themselves are paralleled by the neurodevelopment of areas implicated in executive processing (Davidson et al. 2006, Diamond 2013). Although inhibition improves most rapidly in early childhood, researchers still find plenty of improvement between five and eight years old—for instance, on tasks such as the Day/Night task, which require a child to say "day" upon seeing a picture of the moon, and "night" upon seeing a picture of the sun. Beyond that age, improvement slows significantly although it can still be tracked on complex inhibitory tasks.

This behavioral course mirrors neurodevelopmental findings in the prefrontal cortex (PFC) and PFC-involved network activity, with imaging and EEG results suggesting (1) a transition from diffuse, inefficient activity to focalized, efficient activity during inhibition tasks; (2) the migration of activity frontward and from left-lateralized to right-lateralized, with increasing age and performance; and (3) increased connectivity of frontal brain regions. These structural and functional changes continue well into adolescence (Best, Miller, and Jones 2009). Throughout this period, working memory shows a linear increase across performance measures such as storage, retrieval, speed, and accuracy—sensitive, as with inhibition, to task complexity (Gathercole et al. 2004). This improvement is again paralleled by imaging studies, which show both location and activation-level changes: qualitative change in premotor activation and decreased ventromedial activity, and the shift, prefiguring adolescence, to more frontal regions and anterior cingulate cortex (Bunge and Wright 2007, Scherf, Sweeney, and Luna 2006).

Performance on tasks probing cognitive flexibility (i.e., shifting) continues to improve during this period, as measured by tests such as the Wisconsin Card Sorting Test. In the classic version of that task, the participant must play a "game," sorting cards with different-colored shapes on them by some specified property—for instance, color. After several of these "nonshift" trials, the rules of the game change; the child is instructed to now sort them by shape, say, rather than color. One way that researchers measure performance on such tasks is "shift cost": how much speed or accuracy the participant gives up when moving from nonshift to shift trials. Measured in terms of response time, shift cost was significantly greater for 7- and 11-year-olds than for 15-year-olds (Huizinga, Dolan, and van der Molen 2006). A separate study (Davidson et al. 2006) found that shifting produced decreases in *accuracy* between 9- and 13-year-olds while producing increases in *response time* from 6 years old to adulthood, which the authors see as evidence of an emerging speed/accuracy trade-off—by which children tolerate a slower response if it means a more accurate one—that suggests metacognitive development.[9] Once again, imaging studies line up with these behavioral trajectories, showing activation increasing with age in inferior frontal, parietal, and anterior cingulate regions, core executive areas already implicated in shifting by earlier work with adult brains.

Assessing the planning abilities of children in this age range—as probed by measures such as the Tower of London/Tower of Hanoi tasks—depends heavily on task complexity. As the Tower tasks are manipulated to increase their difficulty, the performance gaps between children and adults, and between different-aged children, start to open up. Planning three moves ahead is present by middle childhood; planning four or five moves ahead does not seem to emerge until late childhood or early adolescence (Best, Miller, and Jones 2009).

Adolescence (13 to 18 Years Old)

Because our view of responsibility is cashed out in terms of the possession of an MWS, and because the PFC is where these functions are for the most part neurally realized, we will need to review the most notable structural and functional changes that have been found to occur in the prefrontal areas of adolescent brains. And while the neuroscientific picture of the brain regions responsible for executive processing has widened, PFC remains the locus of most executive activity—and in terms of structural integrity, connectivity, and function, even older adolescents lack a fully mature PFC (Luciana 2013).

Laurence Steinberg and his colleagues have done some of the most influential work in this area, and they identify four structural developments of particular importance (Monahan, Steinberg, and Piquero 2015): (1) decreases in prefrontal gray matter, mainly in preadolescence and early adolescence, as a result of synaptic pruning; (2) increases in prefrontal white matter, beginning in adolescence and continuing into the 20s, as a result of myelination, the process by which nerve fibers become sheathed in a white, fatty tissue called myelin that fosters more efficient signal transmission in neural circuitry; (3) improvements in connectivity between PFC and other areas, for example, the limbic system, resulting in increased "crosstalk between regions important in the processing of emotional information and those important in self-control" (583); and (4) changes in the density and distribution of dopamine receptors in brain regions connecting the limbic system to prefrontal areas.

These structural changes in the PFC parallel a group of important changes in brain function, detectable across adolescence. Monahan et al. (2015) note three such changes: (1) a "strengthening of activity in brain systems involving self-regulation and self-control," which continues into early adulthood; (2) a host of hormone-related changes to reward response, and (3) increases in the simultaneous involvement of multiple brain regions in response to arousing stimuli (583–4). Some specific executive function (sub)components—the ability to update working memory contents, for one—do seem to reach maturity during adolescence, but many others, such as the ability to inhibit actions and delay gratification or the ability to plan future actions involving other people, continue to change well into early adulthood (Paus 2005, Steinberg et al. 2008). Researchers have also noted a developmental "hitch" in how adolescents' executive functions interact with the brain's emotional systems; adolescents are distinctive in the sensitivity of their executive functions to emotional and reward contexts (Christakou 2014).

Research into the development of executive functions and the brain processes underlying them remains an area of intense and fertile inquiry, and several important questions remain unanswered. Still, there is enough converging evidence from behavioral and neuropsychological research to show that executive functions continue maturing through early adulthood, with a "perfect storm" of high impulsivity, emotional volatility, and reward-seeking behavior occurring during mid-adolescence. And there is little doubt that the unique turbulence of this period at least partly explains the infamous "inverted-U" shape tracking the incidence of criminal and antisocial behavior over the early life span, which begins low, in childhood, peaks during late adolescence, and declines again through the 20s.

Individual performances on a battery of neuropsychological tasks probing executive functions have been found to be uniquely predictive of participants' real-life risky behavior, "above and beyond the variation that was accounted for by personality, age, and sex"—or, for that matter, general intelligence (Pharo et al. 2011, 975). Importantly, it does not seem to be the case that adolescents are disproportionately disposed to risky behaviors because they do not know, or cannot grasp, the dangers their actions court. Rather, they engage in risky behaviors *despite* knowing the risks. This, again, supports the notion that it is adolescents' executive immaturity, especially in the context of high arousal or reward salience, and especially in the presence of their peers, which seems to account most plausibly for their elevated risk of criminal, immoral, or dangerous behavior (Steinberg 2008).

Reviewing the developmental evidence on self-control, Casey (2015) concludes that adolescents have a diminished ability to suppress "inappropriate emotions, desires, and actions in favor of alternative appropriate ones" in the presence of salient environmental cues, to which they are especially sensitive "both behaviorally and neurally" (310). One recent study (Vera-Estay, Dooley, and Beauchamp 2015) found that that executive functions explain a significant amount of the variance found in the maturity of moral reasoning in adolescents, even when controlling for age and intelligence (as measured by IQ). As an example of how executive functions could influence the development of moral reasoning, consider the importance to moral agency of cognitive flexibility, which helps us take others' perspectives and select appropriate rules for behavior in a fluid environment. Explicating their findings, Vera-Estay and colleagues write that "[if] an adolescent's executive functioning is impaired, the quality of both social understanding and MR [moral reasoning] may be affected, resulting in inappropriate social behavior … [that] may become more entrenched when individuals experience rejection from their social group, gravitate toward like-minded peers, and are reinforced for expressing and engaging in negative or antisocial behavior" (Vera-Estay et al. 2015, 28).

Considered together, these findings show juvenile responsibility as a phenomenon under fluid and complex development, different in many significant ways from what we think of as full-blown, adult-level responsibility. Young people find themselves experiencing intense, urgent impulses while lacking the self-regulatory capacities needed to keep from acting on such impulses (Steinberg 2010). Imagine someone turning on a firehose before you've managed to get both your hands on it.

Juveniles also possess less-established characters, defined roughly as sets of stable dispositions to act in certain ways in some relevant set of circumstances.[10] As we discussed in chapter 3, adults are typically able to deliberately intervene on our future selves by forming intentions, training our dispositions, and manipulating our environment, becoming ever more "responsible for who we are" (Roskies 2012, 331). Executive functions such as planning, attentional control, and inhibition enable us not only to make choices for action (given a particular goal), but to make choices establishing future dispositions to act.

Young children largely lack this ability for self-intervention. Parents and other adults initiate and direct its development by pairing positive or negative emotions with particular outcomes and encouraging their children to reflect on the consequences of their choices and devote more time and care to planning their actions. Adolescents begin to exhibit independence regarding habituation of traits and character, but their dispositions have yet to become stable, and their impaired capacity for self-regulation can derail deliberation on the best choice given a desired trait (e.g., courage) or outcome, as well as counteract the pairing of positive emotional reinforcement and good outcomes.

Because juveniles have few stable dispositions and few resources to self-direct the process of habituation, they are especially vulnerable to manipulation of their actions and dispositions. As one example of how neuropsychological immaturity is linked with a heightened susceptibility to external forces of control and influence, a study of ten-year-olds found that the degree of functional connectivity between three different neural systems—including the executive network—correlated with resistance to peer influence (Grosbras et al. 2007). Similar research indicates that early adolescence, compared to late adolescence and adulthood, is marked by a lower ability to resist peer influence (Steinberg and Monahan 2007), a stronger link between delinquent peer association and criminal behavior (Monahan, Steinberg, and Cauffman 2009), and a higher rate of co-offending (Reiss and Farrington 1991, Piquero, Farrington, and Blumstein 2007, Zimring and Laqueur 2015).

Such findings confirm the intuitive sense that, because the young are less able to control their own thoughts and actions, both synchronically and diachronically, they are more susceptible to influence, coercion, and exploitation by others. Children and adolescents are not simply "weaker" versions of adult moral agents, as if they merely had a less developed ability to express a fully formed character through their actions. Rather, their executive immaturity puts them at increased risk of having their agency compromised, broken up, hijacked, or bypassed altogether.

Scalar Responsibility in Juveniles

Purely at the level of common sense—"as any parent knows," in Justice Kennedy's words—it seems quite plausible that children's responsibility for their actions increases in degrees, tracking the improvement of their capacities to grasp and weigh reasons and exercise control over their thoughts, words, and deeds. This commonsense view is reinforced by the broad structure of the criminal law as well as findings from contemporary neuropsychology. Those findings, reviewed here, indicate a complex and gradual development of executive processing in the maturing brains of children and adolescents. As we move from childhood toward adulthood and our executive functions approach maturity, we amass an increasingly rich and varied set of life experiences; little by little we expand our synchronic and diachronic domains of control.

And yet, despite this alignment between science and common sense, philosophers remain hesitant to embrace a scalar conception of responsibility, let alone the specific claim that responsibility "grows" along a gradual trajectory in a normally developing child. To be clear, the idea that *blameworthiness* comes in degrees is fairly widespread in the philosophical literature; the idea that *responsibility* comes in degrees is far more controversial.[11] This strikes us as an odd asymmetry; to take just one high-profile example, Fischer and Ravizza (1998) are happy to endorse a scalar conception of blameworthiness, but their reasons-responsiveness account holds that moral responsibility is a threshold concept: An agent either meets the threshold or does not.[12]

Even if one agrees broadly with reasons-responsiveness accounts like that of Fischer and Ravizza, as we do, it does not seem fully satisfying to only describe these minimum threshold conditions. We agree with Coates and Swenson (2013) that a complete account of responsibility should be sensitive to the intuition that some agents are more responsible than others, and that some agents are systematically more responsible for certain *kinds* of actions than for others. That position echoes H.L.A. Hart's notion of "capacity-responsibility": According to Hart, the capacities necessary to responsibility "may be lacking where there is mental disorder or immaturity ..." and these capacities may be "diminished ... as well as altogether absent" (Hart 1968, 218–28). Our own account is deeply tied to this idea of Hart's; as we argued in chapter 4, capacity responsibility should be understood as comprising an MWS of executive functions. And because we can identify the gradual development toward an MWS in the maturing brain, we contend that juveniles of different ages and levels of cognitive maturity have different levels of responsibility. More specifically, the aggregate weight of neuropsychological evidence[13] indicates that the maturation of juveniles into fully responsible agents closely tracks the development of executive functions. And, we argue, this account of juvenile agency should inform our thinking about when, why, and how to hold juveniles responsible.

We would nevertheless resist the notion that there is some overall quantity of responsibility that one has, determined as a function of one's overall executive abilities. Even if we could track a continuous increase in a juvenile's global executive functioning as he or she approaches adulthood, we should not simply equate these higher levels of executive capacity to higher "amounts" of responsibility *simpliciter*. Instead, we recommend thinking of the scaling up of juvenile responsibility as comprising two interacting factors.

First, a maturing juvenile's increasing degree of responsibility tracks the gradually widening range of situations or circumstances in which his or her developing executive abilities can operate effectively—a function of both executive development itself and the accumulated knowledge gained from experience. Suppose that Tina the teenager is a 15-year-old who, when given a standard battery of "cool" executive function tasks, performs at a high enough level to be considered to have an adult-level set of executive functions. There is nothing farfetched about that prospect, and Tina might fairly be considered fully responsible for choices she makes in "cool" contexts.[14]

At the very least, by our lights, Tina's performance on this battery of tests is prima facie evidence that she is more responsible for what she does than is a 15-year-old who scores significantly worse on those same tests.

But if Tina is, as far as we know, a roughly typical 15-year-old, we would also expect her executive capacities to be systematically prone to failure in contexts of high reward salience or emotional turbulence—situations that tap "hot" executive functions—and we should be accordingly disposed to consider her less responsible, compared to an adult, for actions performed in such contexts. More generally, if Tina commits a crime in circumstances where her adolescent vulnerability to emotional overload and sensation seeking is being exploited, we ought to regard her as comparatively less culpable—not because adolescents are less capable, of conforming their behavior to the law, but because *in that kind of circumstance*, adolescents are distinctively unable to exert adult-level volitional control.

The second factor in our scalar view of juvenile responsibility is that juveniles are responsive to a narrower range of reasons than are adults, on average, and that this range systematically widens as they age toward adulthood. This factor draws on an idea from Coates and Swenson's (2013) attempt to extend a reasons-responsiveness account, in the style of Fischer and Ravizza, to capture degrees of responsibility. Recall that, as we argued in chapter 3, Fischer and Ravizza's theory of responsibility requires that one establish a coherent history of reasons-responsiveness across time. In the language of Coates and Swenson, younger juveniles tend to process morally or legally salient reasons in "less understandable fashion" than do adults. To illustrate this idea, Coates and Swenson employ an example inherited from Fischer and Ravizza (1998): If Larry has very little disposable income, he might understandably see a $1,000 Super Bowl ticket as too pricey. But if Larry did not also see a $500 Super Bowl ticket as too pricey—although less expensive, obviously, than the first ticket—then his reasoning would be less intelligible than that of someone in relevantly similar circumstances who *did* see both tickets as too pricey.

This example is a bit odd, and it does not immediately and neatly transfer to the context of juvenile responsibility—but it is getting at something important. It is surely intelligible, for instance, that 15-year-old Tina would value peer approval very highly. Of course, there are dangers or harms that Tina would never risk even for a huge boost in popularity—suppose that she would obey the old parental saw about not jumping off a bridge, even if all her friends did the same—but we might still expect Tina to be especially prone to more risky behavior when peer approval is at stake because Tina's mechanisms for processing those kinds of reasons simply do not operate at adult levels of intelligibility. And this disposition might manifest in behaviors that risk harms or wrongs in the pursuit of peer approval, of the sort that would be unintelligible (or less so) if performed by an adult. This expectation is not only a commonsense one; as we have shown, it is supported by a wealth of evidence from the study of executive functioning in adolescent brains. To be clear, we are not saying that Tina is unable to grasp how salient the reasons against her risky behaviors are. As the data reviewed here show, teenagers do risky things even though they understand the risks. Tina's underdeveloped responsiveness to moral or legal reasons is (mostly) not a matter of being unable to recognize those reasons; Tina is probably about as competent as an adult when it comes to accurately judging what sorts of actions would be dangerous, foolish, illegal, or wrong. Where Tina is most likely to fall short is in her ability to, for instance, inhibit actions she has accurately judged to be dangerous, foolish, illegal, or wrong.

We can now recapitulate our two-factor view of why juveniles, as their executive processes mature, possess gradually increasing degrees of responsibility. The first factor is the widening range of *circumstances* in which a juvenile's executive capacities operate effectively, and the second factor is the widening range of *reasons* to which juveniles are intelligibly responsive (the value of a long-term benefit that requires sacrifice of a short-term, emotionally salient benefit would be one obvious example of the sort of reason to which juveniles respond in increasingly intelligible ways). These factors depend upon the development of an MWS, the components of which emerge gradually and along distinct but interdependent trajectories. Developmental neuropsychology gives us the tools and the data to describe how those components mature over the first 25 years of life, and that maturation explains how we go from being infants with no meaningful degree of responsibility to being adults with full-blown freedom and responsibility.

Consider the following remark from Brink and Nelkin (2013): "[B]eing a responsible agent," they write, "is not merely having the capacity to tell right from wrong but also the capacity to regulate one's actions in accordance with this normative knowledge. This kind of volitional capacity requires emotional and appetitive capacities to enable one to form intentions based on one's optimizing judgments and execute these intentions over time, despite distraction and temptation" (Brink and Nelkin 2013, 297). As we discussed in chapter 4, we find this picture of responsible agency quite attractive, particularly in its emphasis on both the cognitive and the underappreciated volitional components of responsibility. However, we think Brink and Nelkin risk a misleading oversimplification in emphasizing the "emotional and appetitive" capacities enabling volitional control. Forming intentions "based on … optimizing judgments" and being able to "execute these intentions over time, despite distraction and temptation"—these are precisely the forms of volitional control that fall under the aegis of the executive functions, which are still maturing significantly well into adolescence.

A Graded Structure of Legal Responsibility

We have offered an empirically plausible view of how executive capacities develop from infancy through adulthood and of how that development underwrites the gradual emergence of full-fledged responsibility. How might we apply this scalar view to the practice of juvenile justice? The question presents an immediate and obvious obstacle: The criminal law requires tractable standards and bright lines that can be applied with relative uniformity across widely varying cases, and that practical need exists in unavoidable tension with the gradual and individually variable nature of neurobiological maturation. Different executive components mature along separable developmental trajectories, as do the brain systems subserving those functions, and there can be wide disparities between individual juveniles when it comes to executive maturity. There is obviously no single age at which a child becomes, all at once, a "normatively competent" adult (Brink 2004, Scott and Steinberg 2009, Steinberg 2012). Yet, as David Brink points out, "[T]he law often draws lines in ways that generally, but nonetheless imperfectly, track the facts that matter," and the fact that such imperfections are inevitable should not stop us from trying, as best we can, to "achieve individualized justice, consistent with the use of a generally but imperfectly reliable boundary marker" (Brink 2004, 1578).

With that principle in mind, we will draw some lines of our own, by recommending a graded structure of culpability for juveniles. Rather than a single cutoff separating juveniles from adults, we propose a sequence of age ranges during which juveniles' executive functions tend to achieve maturity. Such a structure, we believe, is the best choice for balancing the practical demands of criminal jurisprudence with a scientifically and philosophically defensible view of developing responsibility. In a previous article (Fagan, Hirstein, and Sifferd 2016), we recommended a staged structure of culpability for child soldiers under international criminal law. We endorse such a structure because we think it has enough uniformity, rooted in broadly stable facts about executive development, to be consistently and fairly applied across a wide range of contexts. At the same time, it has enough flexibility to accommodate the context-specific facts of particular cases and local norms and to mark the gradual, individually variable maturation of executive functioning that occurs in late adolescence and early adulthood.

We suggest that, in the guilt phase of juvenile justice, a set of legal categories be established that more accurately captures the gradual development of adult-level executive capacities:

- Age 12 and under: an irrebuttable presumption of incapacity on grounds of immaturity.
- Ages 13 to 17: an irrebuttable presumption of diminished capacity on grounds of immaturity.
- Ages 18 to 21: a rebuttable presumption of diminished capacity on grounds of immaturity. That is, evidence could in principle be produced to overturn this presumption, but it would have to be good enough to rationally compel a reasonable person to think the presumption false.
- Ages 22 and up: a presumption of adult-level capacity that cannot be rebutted *on grounds of immaturity*. (Obviously other paths to a mental incapacity defense would remain open, but not one that leaned solely or primarily on the immaturity of the offender.)

Although this proposal would represent a radical change from the status quo, there are models within the American justice system for how it might be implemented. In 2015, the city of San Francisco established "Young Adult Court" to handle eligible offenders from ages 18 to 25. On its website, the Court explicitly cites contemporary neuropsychological evidence to support its position that young adults "are qualitatively different in development, skills, and needs from both children and older adults." By carving out space for a category distinct from the traditional juvenile–adult binary, the Court (2017) aims to "align opportunities for accountability and transformation with the unique needs and developmental stage of this age group."

Experiments like San Francisco's suggest that a multitiered scheme of age-based legal categories may provide a balance between "individualized justice," in Brink's phrase, and the need for practical usefulness. However, if this kind of graded framework seems too radical a change—or even if it does not—we would favor an age-of-adulthood cutoff that more closely matches our best understanding of when full executive maturity tends to be achieved. This would entail raising the legal age of adulthood for the purposes of criminal prosecution to 21, so that all offenders under 21 are handled first by the juvenile justice system, and tightening the restrictions governing the practice of juvenile transfer. The rules of transfer should be based primarily upon an assessment of the offender's cognitive capacities, rather than indexed to the severity of offense.

Because our focus in this book is on the neuroscientific basis of legal responsibility, our suggestions for rethinking juvenile justice have mostly aimed at the guilt phase of that system. But even after guilt is established, the reduced executive capacity of juvenile offenders should be reflected in decisions made at the sentencing phase. This is true, we would argue, regardless of one's position on the primary function of criminal punishment. From a retributivist perspective, an accurate view of how executive functions develop demands that we take seriously the reduced culpability and therefore blameworthiness of juvenile offenders, which in turn should dispose us toward less harsh forms of punishment. And from a forward-looking view of punishment's functions, the high recidivism rates among juvenile offenders, especially those sent to adult prison, suggest that our present practices are poorly serving the aims of deterrence and rehabilitation. Juveniles serving sentences in adult prisons are more likely to recidivate once released for many reasons: With a felony record they are less likely to find work, and they suffer from increased alienation from their family and other supportive relationships. We should work toward intervention strategies specifically intended to promote the development of executive capacities—such measures may include skills and job training, educational programming, mindfulness programs, yoga, chess, gardening, and exercise programs—by which we can help juvenile offenders become more deserving of full-fledged ascriptions of responsibility and better at behaving lawfully.[15] We discuss these programs that may protect, instead of diminish, moral and legal agency in detail with regard to adult criminal sentences in chapter 10.

Responsibility and the Aging Brain

The foregoing discussion, tracking the development of full-blown legal responsibility as we age from children into adults, raises the question of whether a mirror image of this trajectory might be found at the end of the life span. Cognitive abilities decrease with advanced age (Cohen-Mansfield et al. 2013), and the study of mental decline in older adults has advanced enough to suggest that cognitive and volitional capacities, even in a healthy aging brain, could deteriorate badly enough to erode one's responsibility. If the ability to inhibit action becomes sufficiently degraded, for example, an older person might lack the self-control to keep from engaging in a behavior that, in younger days, that person would have been able to avoid. Indeed, some recent studies of aging populations have found a robust relationship between self-control and the likelihood of engaging in dangerous or risky behavior, including criminal offending and prescription drug misuse (Holtfreter, Reisig, and O'Neal 2015, Wolfe, Reisig, and Holtfreter 2016, Hirtenlehner and Kunz 2016).

We have argued in this chapter that juveniles—even older ones—may fall short of an MWS of executive functions, and that legal practices of assigning culpability to juveniles should reflect their executive immaturity. And if the executive "profile" of juveniles can be roughly compared to that of older adults, then by parity of reasoning we ought to explore a legal approach meant to accommodate the diminished capacity found in the latter population. The analogy with juveniles seems promising. A host of studies have shown that, even in the absence of dementia, executive processes are "particularly sensitive" to age-related decline in older adults (Manard et al. 2016, 563). Researchers have found age-related declines across all components of the Miyake-style model of executive functioning (Manard et al. 2016, Rozas, Juncos-Rabadán, and González 2008, Hedden and Gabrieli 2004). And while a number of specific neuropsychological models have been proposed to explain these declines, meta-analyses seem clearly to establish that, in general, the PFC is more significantly and deleteriously affected by aging, even compared to other neocortical regions (Raz and Rodrigue 2006, Manard et al. 2016).

We are suggesting neither that older adults are equivalent to juveniles with respect to executive function or capacity responsibility, nor that both populations can be put under a single legal category, since a number of obvious and relevant differences exist between them. Still, we think there are sufficient prima facie grounds for seriously questioning the way Anglo-American criminal law assigns blame and punishment to aging adults. Some degree of symmetry appears to exist between the executive deficiencies found in developing and aging brains, but that symmetry is scarcely reflected in the criminal law. An entire edifice of statutes and institutions has been established to handle juvenile defendants, but there is no comparable structure in place for defendants in the later years of life. Anglo-American law is certainly aware that older adults can exhibit reduced capacity for decision-making, but that fact most frequently shows up in contexts of legal guardianship or consent to medical treatment. When advanced age *does* enter into a legal proceeding that bears on criminal punishment, it tends to happen at the sentencing stage—for instance, when a judge weighs the costs and benefits of imprisoning a defendant with advanced dementia.

The question of whether older adults—in the absence of a specific mental disorder—might have a claim to diminished criminal responsibility has not received the attention it deserves. Nor has enough attention been paid to aging offenders who may have suffered enough executive decline to call into question whether they even understand the reason for their continued punishment. Even if they were fully responsible at the time of their crimes, one might think the purposes of criminal justice are ill served by punishing offenders who do not understand why they are being punished. These issues are important directions for future research, especially as America's massive prison population grows older, and we are hopeful that our executive account of responsibility might be plausibly extended to cover aging populations.

9 Responsibility, Capacity, and Insanity

We have thus far argued that executive functions ground the forms of responsibility employed in our everyday assessments (chapter 3) as well as the criminal law (chapter 4)—and, more specifically, that capacity responsibility requires what we call an MWS (minimal working set of executive functions). Possessing an MWS is necessary for the "normative competence" that the criminal law considers requisite for assignments of criminal culpability (Brink and Nelkin 2013); it enables sensitivity to both morally or legally relevant reasons and the volitional control to act in accordance with those reasons.

The previous chapter explored one prominent category of offenders often considered exempt from or subject to diminished legal responsibility, juveniles. In this chapter we will consider the class of offenders who potentially fail to satisfy capacity responsibility on grounds of mental illness. Legal insanity is a concept familiar to any criminal attorney—or anyone who has watched a handful of *Law & Order* episodes, for that matter—but the philosophical and scientific foundations of the insanity defense are still muddled, and controversy about its application still exists in public, academic, and legal circles. A number of recent high-profile cases prominently featuring insanity defenses, such as that of Anders Breivik in Norway, whose story we told in chapter 1, have been driving this controversy, raising questions about the so-called medical model of legal insanity: namely, whether the incapacities that qualify a defendant as legally insane need be the result of a mental disorder or disease and, if so, why (Bortolotti, Broome, and Mameli 2014, Moore 2014).

We will try to clear up some of this confusion in this chapter. Our executive theory of responsibility can explain and broadly justify the insanity defense's role and application while also suggesting ways it can be usefully reformed. For example, American legal jurisdictions remain split between those using the M'Naghten standard of legal insanity and those using the standard recommended in the American Law Institute's Model Penal Code. If we are right in arguing that legal responsibility is grounded in executive capacities, then the MPC rule—which includes both cognitive and volitional conditions—is preferable to the M'Naghten rule, which uses only a cognitive test. Our theory can also shed light on the question of what relevance, if any, exists between clinical judgments of mental illness and legal judgments of insanity or incapacity. We will then take up the question of whether psychopaths should qualify as exempt from criminal responsibility by the lights of our theory, or at least as eligible for diminished-capacity status.

Two Standards of Legal Insanity

On the model of legal responsibility that we have endorsed, recognizing that one's act is immoral or illegal is not always enough to make someone refrain from unlawful behavior. If a defendant recognizes some act as forbidden by the criminal law but cannot avoid such conduct even given this recognition, it does not further the purposes of the criminal law to hold the defendant responsible for violating the law.[1] This is true regardless of the reasons behind the defendant's incapacity to conform his or her behavior to the law (with some rare and rather outlandish exceptions). If Sue is coerced via a gun to the head to steal another's bag, she ought to be excused, and if she has a mental disorder rendering her incapable of resisting an impulse or desire to steal the bag, she ought to similarly be excused. She is incapable of being deterred from stealing the bag, and she does not deserve retribution for stealing it.

It is therefore unfortunate that in many U.S. jurisdictions, a defendant who recognized his or her act as criminal or wrong but lacked the capacity to inhibit an act based upon this knowledge is deemed fully culpable for that act under the criminal law. Such jurisdictions adhere to the M'Naghten test for legal insanity, which excuses a defendant who, because of a severe mental disease or defect, is unable to appreciate the nature and quality of the wrongfulness of his or her act. A purely cognitive standard, lacking any mention of volitional or control-based factors, the M'Naghten test still applies in the majority of U.S. states and is reflected in the federal insanity test, as well as a number of other common-law jurisdictions. The MPC, which requires that a defendant lack "substantial capacity to either appreciate the criminality of his conduct or to conform his conduct to the requirements of the law" (MPC 4.01), remains the minority standard (Penney 2012).[2]

David Brink and Dana Nelkin provide an excellent critique of purely cognitive tests of legal insanity that have emerged in the United States and elsewhere. They argue that criminal responsibility requires a "fair opportunity to avoid wrongdoing," and that this fair opportunity to be law-abiding requires both normative competence and situational control (Brink and Nelkin 2013, 284). If a justificatory defense is not available, then a person can only avoid culpability by appeal to an excuse: that he or she lacked either normative competence or situational control at the time he or she committed a crime. If this is the right way to think about such categories of legal excuse, then the insanity defense rests upon either a denial of a defendant's ability to recognize norms or legal rules or a denial of the defendant's ability to conform his or her behavior to such rules (Brink 2013, 3–5). As we mentioned in chapter 4, Brink and Nelkin claim that each of the two requirements is independently necessary for normative competence: Neither actors lacking the cognitive capacity to recognize or appreciate legal or moral norms nor actors lacking the volitional capacity required to conform conduct to this normative knowledge are responsible under the law. Because of this, they support the MPC's test for legal insanity over the traditional common-law approach, which utilizes the M'Naghten rule. In chapter 4 we also noted that Brink and Nelkin argue for a fairly subtle understanding of the capacity to recognize wrongdoing: It not only involves the capacity to know whether an action or its result violates a moral or legal norm, but to understand the nature of one's act in relation to the law, as well as in relation to the circumstances surrounding the act, and sufficient volitional control to refrain from performing an act one has judged to be immoral or illegal (Brink and Nelkin 2013).

Despite the arguments made by Brink and Nelkin on behalf of a "control" element to normative competence, legal scholars have remained skeptical about the MPC's inclusion of volitional capacity as excusatory. Some of these skeptics, as Steven Penney (2012) suggests, doubt the very idea of a person who could appreciate the wrongfulness of his or her act but could not refrain from acting wrongfully. For example, Stephen Morse has argued that the notion of an irresistible urge is nonsensical because it is so difficult to make sense of what it means for an urge to be so strong it is "irresistible" as opposed to urges that could, under some conditions (such as threat of death), be resisted (Morse 2002). It seems, says Morse, that persons under sufficient threat of sanction can resist *any* urge. This is the case even for those with mental conditions such as obsessive-compulsive disorder: A man obsessively locking and relocking his door can stop himself from the behavior if his life is threatened because his house is on fire.

Brink and Nelkin claim, however, that resistibility is a perfectly coherent concept—a "modal notion," as they put it. How "unconquerable or uncircumventable" an impulse must be to qualify as excusing, and how we can distinguish the inability to overcome and the failure to overcome—these questions raise interesting pragmatic and epistemic concerns. But there is no conceptual bar to the law's drawing a line here, just as it does when it determines what blood alcohol level makes one legally drunk or what age makes one old enough to vote (Brink and Nelkin 2013, 300). Cases in which a person would only act differently under a threat of imminent death, for example, seem to be obvious cases in which we should excuse the person. But this cannot be the standard, Brink and Nelkin argue, or the threshold of excuse would be too difficult to meet (Brink and Nelkin 2013). Rather than requiring complete incapacity to resist, as M'Naghten does, the MPC seems correct in requiring a legally insane defendant lack *substantial* capacity (Brink 2013, 12–16). Putting this point in the language of Fischer and Ravizza's reasons-responsiveness theory, Brink and Nelkin see the MPC's standard as requiring normatively competent agents to be moderately, and not just weakly, reasons-responsive.[3] Someone who was only weakly reasons-responsive at the time of his or her offense was likely unable to both recognize wrongdoing *and* refrain from acting in response to this recognition. If such a person's lack of substantial capacity is attributable to a mental illness, that person may be rightfully deemed legally insane.

Our approach supports Brink and Nelkin's "fair opportunity" theory and, accordingly, the MPC standard: Legal sanity requires a certain degree of both cognitive and volitional competence, which we cash out in terms of the possession of an MWS. Only persons with such capacities can recognize their acts as immoral or illegal *and* act or refrain from acting based upon this recognition. A person lacking such capacities is not deserving of blame or retribution, nor are consequentialist purposes for criminal punishment served by holding a person lacking these capacities responsible. It is futile to threaten a person with punishment who cannot be swayed.

Other skeptics about a volitional component of legal insanity, while allowing the in-principle possibility of volitional incapacity, doubt that neuroscientific and psychological tools can tell us enough about this phenomenon for "the law of criminal responsibility [to] practically accommodate" it (Penny 2012, 99). As the next section shows, however, volitional impairments are indeed a genuine phenomenon in the forms of mental illness most commonly associated with legal insanity pleas, and the tools of neuroscience and psychology can tell us much about those impairments. We believe that a standard of legal insanity, carefully construed and applied, can practically accommodate evidence of executive dysfunction in persons with mental illness.

Executive Dysfunction and Legal Insanity

If we are right that executive functions underlie the cognitive and volitional competencies picked out by the MPC, then justifiable claims of legal insanity should track significant deficits in executive functioning. To see whether this is true, we will review some evidence of systemic executive dysfunction in persons with mental illness of the sorts that have tended to ground successful insanity defenses. Despite the pop-cultural ubiquity of the insanity defense, it is quite rare for a defendant to be excused from criminal responsibility on grounds of legal insanity—and in most of those cases, the defendant was a person with schizophrenia (or, less frequently, bipolar disorder).

Schizophrenia is perhaps most commonly associated with the "positive" symptoms of hallucinations, delusions, and bizarre-sounding beliefs. Less widely appreciated—but crucial to questions of responsibility—is the fact that persons with schizophrenia often have serious deficits falling across multiple areas of executive function. They show decreased cognitive processing speed, easily depleted sequencing capacity, difficulty in multitasking, and inability to keep from jumping to conclusions, along with sometimes severe deficiencies in attentional set shifting, planning and strategy capacity, and online use of working memory.[4]

In one study, participants were both given cognitive testing and judged based on their ability to perform a set of activities involved in daily living;

they had to choose a menu for dinner, pick up needed items from a store, and then prepare a meal (Semkovska et al. 2004). Subjects with schizophrenia had significant deficits in the ability to prepare a meal with multiple items at the same time—they could initiate the process, but they could not complete it. In another study, participants were asked to use a computerized meeting preparation task to organize a meeting. The ability of people with schizophrenia to successfully complete the exercise was significantly lower than normal, and it took them longer to finish the task (Laloyaux et al. 2014).

We should note two further points about the patterns of executive dysfunction in persons with schizophrenia. First, these behavioral data are consistent with neurological studies of schizophrenic patients. For example, one study found patients with schizophrenia demonstrated selective deficits in cortical inhibition of gamma-band oscillations in the dorsolateral prefrontal cortex compared to those with bipolar disorder and healthy patients, suggesting that a frontally based neurophysiological deficit could be responsible for the spectrum of cognitive control deficits associated with schizophrenia (Radhu et al. 2015). Second, although schizophrenic patients' positive psychotic symptoms often fluctuate, these executive deficits have been found *not* to track the severity of those symptoms; they persist even during periods of remission (Harvey et al. 2006, Krishnadas et al. 2014).

Such executive deficits, if they are sufficiently severe and widespread, could on their own exculpate unlawful behavior. But it is especially telling to consider how those deficits might interact with the hallucinatory or delusive beliefs that sometimes afflict schizophrenic patients. One particularly relevant recent study used a task designed to measure schizophrenic subjects' proneness to jump to conclusions. Subjects had to guess which of two jars a sequence of balls had come from; before guessing, they were allowed to draw as many balls as they liked from each jar—the fewer balls before guessing, in general, the stronger the hypothesized disposition to jump to conclusions. These tests showed that schizophrenics have a heightened tendency to jump to conclusions; indeed, the higher the deficit in overall executive function in schizophrenics, the higher the tendency to jump to conclusions (Ochoa et al. 2014). This is troubling on its own, but think of how that tendency could exacerbate a schizophrenic person's risk of engaging in harmful or risky behavior: Such a person might suffer from

(1) repeated and persistent hallucinations, (2) a degraded ability to recognize his or her hallucinations as unsupported by adequate evidence, and (3) a heightened tendency to act in accordance with hallucinations. As we noted in chapter 2, executive processes typically enable us to reject spurious perceptions and memories. We depend on them to correct for distorted perceptions and to prevent distorted perceptions from being coined into beliefs. Such cognitive control is crucial to mental competence; what makes a person who is experiencing hallucinations mentally incompetent is not the presence of hallucinations, but the inability to avoid taking them for reality.

We all experience strange thoughts on occasion: the feeling that someone is watching us, the odd notion that we had a causal influence on something where no reasonable physical explanation is available (e.g., I flip a light switch and a car horn honks outside), the idea that someone might be plotting against us. Executive functions are required in order to assess the plausibility of a thought. They accomplish the comparisons with other beliefs and with memories needed for the assessment. A person with delusions will only act on the basis of those delusions if he or she persists in believing them to be true. The formation of a persistent delusion thus has at least two levels. First, a spurious thought is created, typically due to compromised perceptual or mnemonic faculties, but sometimes also traceable to the emotional systems. Second, there is an executive failure to properly assess the thought and reject it.[5]

We posit that the schizophrenic's substantial executive deficits, working together with a tendency toward hallucinations, delusions, paranoia, and other heightened emotional states, can undermine the schizophrenic's culpable agency. Schizophrenia, and possibly other mental illnesses including bipolar disorder, may result in a lowered executive capacity to identify and correct for hallucinations, delusions, extreme paranoia, or fear and anger, and to inhibit action based upon these ill-grounded beliefs. Further, schizophrenic persons' executive deficits make them more likely to jump to hasty conclusions regarding their hallucinations or delusions. And even if they are capable of momentary correction, they may be more likely to lose attentional focus on the correct information or perception, and lose the ability to inhibit wrongful behavior.

Further, while the evidence regarding mental illness and executive function indicates that while executive functions tend to cluster, such that a deficit in one often occurs alongside other deficits, they are separable in certain tasks and contexts (Garavan et al. 2002, Miyake and Friedman 2012). It is at least in theory possible that a mental illness could keep a defendant's capacity to recognize wrongdoing intact but almost remove the capacity for inhibition. It is possible that mentally ill defendants could, for example, recognize that the devil telling them to set fire to the house is a hallucination but feel such strong emotions in response to the hallucination that they cannot inhibit the action driven by their fear. It may even be the case that such defendants would set fire to the house "with a policeman at their elbow."[6] And although "no single diagnostic or evaluative tool ... can establish whether a defendant was capable of control at the relevant moment," clinicians using these tools in combination "are able to measure control as accurately as cognitive impairment" (Penney 2012, 101). The evidence reviewed above, we believe, shows that certain mentally disordered persons may fail to satisfy the volitional condition of normative competence *even if* they satisfy the cognitive condition, and that clinical evaluators can give reliable evidence of that failure. Thus, a two-factor standard of legal insanity—one including both cognitive and volitional factors—is necessary. If a defendant's mental illness or brain abnormality renders that defendant unable to inhibit an act even while grasping its wrongfulness, one can hardly claim that defendant had a fair opportunity to avoid wrongdoing. The legal concept of insanity should reflect this fact.

The Relevance of Mental Illness to Legal Insanity: Moore's Challenge

Standing apart from the issue of whether justice demands a legal insanity standard with both cognitive and volitional factors, there is a crucial question that any plausible test for legal insanity must answer: What, if any, is the precise relevance of a clinically diagnosed mental illness to a legal judgment of insanity? After all, the M'Naghten and MPC standards both require not only that a legally insane defendant lack certain competences, but that this lack be the result of a mental "disease or defect." Legal insanity is, obviously, a legal concept, with a history and purpose defined by the aims of criminal justice—and just as obviously, "mental disease," "mental illness," and similar terms are medical or medico-scientific concepts, with their own distinctive origins and purposes. How should we understand the relation between these concepts?

One way to put this question, as Michael Moore (2014) has done, is to ask whether the extant tests of legal insanity assume that a mental illness or defect is *strongly* relevant to these tests' notion of responsibility or only *weakly* relevant. These terms, as Moore uses them, have precise technical meanings: A medical concept is strongly relevant if it, "by itself, captures conditions of moral and legal excuse" and weakly relevant if it can only capture such conditions when "conjoined with other non-medical concepts" (Moore 2014, 14). Considered in such terms, the M'Naghten and MPC tests seem to assume a relation of weak relevance between mental illness and legal insanity. These standards hold that a mental disease or defect is not, by itself, sufficient to excuse someone from legal responsibility: The disease or defect must cause substantial cognitive or volitional incapacity. But one might wonder, if this substantial incapacity is doing the exculpatory work, why the *cause* of that incapacity should matter to determining legal responsibility. If being unable to control one's behavior, for instance, is thought to be generally sufficient for excusing one from legal responsibility, then why does it matter whether such inability is caused by mental illness rather than something else?

We are approaching the horns of a dilemma that, according to Moore, bedevils any weak-relevance test, all of which share the view that "mental illness is relevant to responsibility only when it either causes the existence of (or is at least accompanied by) some Factor X" (Moore 2014, 17). The weak-relevance theorist, argues Moore, occupies an untenable position: Her view either collapses into a position asserting *no* relevance between mental illness and responsibility (the first horn) or else must retreat to the relative safety of the strong-relevance position (the second horn). One might reconstruct Moore's argument as follows.

If Weak Relevance (WR) holds that "[m]ental illness is relevant to responsibility only when it either causes the existence of (or is at least accompanied by) some Factor X," then any Factor X on offer—Moore has four sorts in mind—is either independently exculpatory or not (Moore 2014, 17). If Factor X is independently exculpatory (e.g., "lack of substantial capacity to conform [one's] conduct to the requirements of law"), then it excuses the person whether or not it results from (or accompanies) mental illness, in which case mental illness is irrelevant to responsibility. If Factor X isn't independently exculpatory (e.g., ignorance of law), then mental illness is doing *all* the exculpatory work, in which case mental illness is exculpatory whether or not it causes or is accompanied by Factor X. But this is just the *strong*-relevance position, which is obviously at odds with WR. So, since WR turns out to be false whether or not Factor X is independently exculpatory, WR is false.

Because he advocates a thesis of strong relevance between the medical concept of psychosis and the excusatory condition of legal insanity, Moore himself does not have to face this dilemma. Yet his remarks may contain the basis of a plausible answer to it. Moore, quoting Joel Feinberg (1970), argues that there is something "'special' about the way mental illness excuses, when it does excuse. Its specialness consists in its not needing the excusing force of other conditions (such as lack of voluntariness of action, etc.); it does its excusing work all by itself" (Moore 2014, 44). The illness does not (or need not) excuse someone by causing the criminal act; rather, it is a status defense. Mental illness excuses individuals, when it does, by denying responsible agency to the legally insane, such that the law does not apply to them: Like very young children and animals, they are not subject to its dictates. The medical term "psychosis" is a serviceable proxy for the legal concept of insanity, Moore argues, because it designates a person who is very seriously mentally ill.

Moore is right that mental illness is special in its exculpatory power, but he would, of course, admit that not just *any* mental illness is inherently exculpatory. In general, that a defendant suffered from depression or a phobia is likely to be irrelevant to his or her criminal culpability. We claim that a defendant's mental illness excuses the defendant from responsibility, when it does, precisely because it has prevented the defendant from having a fair opportunity to avoid wrongdoing by undermining either the cognitive or volitional elements of normative competence. Therefore, for a mental illness to excuse a defendant in the way required by the insanity defense, it must be of the sort and severity that substantially degrades one's executive function. On the other hand, executive dysfunction is not inherently exculpatory either—it must be dysfunction of the sort that results from mental illness. If a defendant's executive deficits could be traced back to his or her own diachronic agency, they ought not to excuse the defendant, for the same sorts of reasons that that self-induced intoxication or duress ought not to excuse the defendant (as we saw in the case of alcoholic blackout in chapter 7). One who has intentionally cultivated the trait of being hasty and quick to anger—training oneself to suppress or bypass a natural tendency to inhibit violent or aggressive behavior—should not be excused for a crime, even if it can be shown that the crime resulted from an executive deficit. Of course, it will be extremely rare that one's executively guided choices result in the level of executive damage needed for legal insanity. A person who continues to abuse alcohol, despite being fully aware of a worsening susceptibility to alcohol-induced psychosis, would perhaps qualify as an example.

By viewing legal insanity through the lens of executive dysfunction, then, we can plausibly maintain a weak-relevance position. Our account avoids the first horn of Moore's dilemma: By offering some mental illnesses as a special form of executive dysfunction, it preserves the claim that such mental illnesses are specially relevant to legal insanity. And it also avoids the second horn: By making executive dysfunction the bridge between mental illness and legal insanity, it preserves the insight that even persons suffering from certain severe forms of mental illness may not be excused as legally insane, so long as that illness has left their executive functions essentially intact.[7]

Our position is not only capable of avoiding Moore's dilemma while retaining his core insight about the exculpatory specialness of mental illness; it also moves past Moore's problematic endorsement of the quasi-medical term "psychosis" as a proxy for terms like "legally insane" or "so mentally ill as to be nonresponsible." According to Moore, it is a bit of "interdisciplinary serendipity" that the psychiatric term "psychosis" maps neatly onto the legal term "insanity" (Moore 2014, 45). However, this serendipity is no mere stroke of luck. As we have argued here and elsewhere, criminal legal doctrines, as systematized repositories of folk-psychological beliefs, have long been aiming at a target class of conditions that have only recently become amenable to fine-grained analysis using the terms and concepts of science and medicine (Sifferd 2006, Hirstein and Sifferd 2011). What Moore and the folk have in mind when they say things like "seriously mental ill" or "crazy," in the context of legal or moral judgments, are just those types and severity levels of mental illness that substantially disable executive functioning, thereby producing the very cognitive and volitional incapacities the MPC test is designed to probe.

The Case of Anders Breivik

The Norwegian case of 2011 mass killer Anders Breivik—recounted in the first chapter of this book—has drawn significant academic interest from legal scholars, philosophers, and psychiatric experts. Indeed, Moore's challenge is made in an article prompted by the Breivik case. After killing 77 people—69 of them participants in a youth summer camp—Breivik claimed his actions were intended to draw attention to a compendium of far-right militant texts he distributed on the day of the attacks (Pidd 2012, Flynn 2012, Knausgaard 2015). An initial court-ordered psychiatric evaluation resulted in a diagnosis of paranoid schizophrenia, in part due to Breivik's having persistent, systematized, bizarre delusions, including a belief that he was a member of a Knights Templar organization (one that, according to Norwegian police, does not exist).

Norway's criminal code stipulates that any defendant who evinces psychotic symptoms at the time of his crime is not criminally responsible by reason of insanity (Bortolotti et al. 2014). Thus, Norway's test for legal insanity qualifies, in Moore's (2014) proposed framework, as a strong-relevance view: Mental illness is a status defense, where any defendant diagnosed as having a certain mental illness at the time of the crime is excused. The definition of psychosis, according to Norwegian statute, is to be determined by current psychiatric diagnostic manuals. The *Diagnostic and Statistical Manual of Mental Disorders* (4th ed., *DSM-IV*), the most prominent diagnostic manual at the time of Breivik's evaluations, defines psychotic symptoms as involving the presence of hallucinations or delusions (1994). Because Breivik's initial evaluation resulted in a diagnosis that included psychotic symptoms, it was widely thought that he might be excused from criminal responsibility. The prospect of Breivik escaping punishment on grounds of legal insanity caused widespread outrage in Norway; a second psychiatric evaluation was ordered, which judged Breivik not to have been psychotic at the time of the attacks, diagnosing him instead with narcissistic personality disorder. In the end, because of this second psychiatric evaluation, Breivik was found legally sane and guilty of 77 counts of murder. He was given the maximum criminal sentence allowable in Norway: 21 years, with the possibility of an extension if he is deemed to be a danger to society. Breivik will probably be incarcerated for the rest of his life.

In an analysis of Breivik's case, Bortolotti, Broome, and Mameli (2014) claim that even if Breivik did have delusions, and thus was properly diagnosed as psychotic at the time of his crime, this fact may be completely irrelevant to whether Breivik is criminally responsible for his acts. Indeed, they argue, *no* particular set of psychiatric symptoms or diagnoses is sufficient to actually make a defendant legally insane. Because the level of cognitive and social functioning can vary widely among those with psychosis and schizophrenia, they argue, a diagnosis or the symptom of hallucinations or delusions itself does not necessarily indicate the quality of a persons' legal and moral agency.

As already indicated above, we completely agree with this assessment. Someone with an MWS may be capable of detecting problems with hallucinations and inhibiting behavior based on them. Yet we disagree with Bortolotti et al. (2014) in their suggestion that the presence of mental illness is irrelevant to a determination of legal insanity. A diagnosis of schizophrenia or psychosis may well include an assessment of severe executive deficits, and therefore such diagnoses can figure in a claim of compromised legal agency. And importantly, such a diagnosis will often identify executive deficits that are not the fault of the person diagnosed. We certainly concede that a diagnosis of schizophrenia may not indicate serious executive dysfunction; that is why such a diagnosis, while potentially relevant to a claim of legal insanity, is an unreliable stand-in for a more direct determination that the defendant is legally insane because of executive deficits. Indeed, that is partly why we favor the MPC test for legal insanity: That test asks the court, in folk-psychological terms, to look for cognitive and volitional deficits in the defendant's executive functions related to a mental illness. Experts may then offer evidence regarding the defendant's diagnoses and whether the cognitive and volitional deficits suffered by the defendant are substantial enough to warrant an excuse.

Bortolotti et al. are on the right track when they claim that, if Breivik's "poor reality testing (or some other relevant cognitive deficit associated with delusion formation) is affecting the beliefs he is prepared to endorse to the extent that such beliefs are implausible even to members of groups inclined to share his political and ideological views, then maybe such failure of reality testing (or other relevant cognitive deficit) is also implicated in his decision-making processes, including those processes that led him to his criminal acts" (Bortolotti et al. 2014, 380). What matters to responsibility is not the presence of false or bizarre beliefs themselves, but the extent to which a defendant has the capacity to identify and reject them or inhibit behavior based upon them. Sane persons may have bizarre beliefs, delusions, or even hallucinations from time to time (e.g., "I swear I saw a scary looking old man sitting over there!" when it is really just a tree, or "That person walking behind me must be out to get me!"). Executive functions shape how one *responds* to these delusions or hallucinations, though; those responses reveal the difference between legal sanity and insanity. Walking on an otherwise empty street at night, I may be plagued by the feeling that the little old lady walking behind me intends to do me some evil; I may even entertain the notion that she is only pretending to be a little old lady. If my executive processes are functioning normally, I might respond to this unsettling situation by turning to ask her if anything is wrong or trying to determine whether she is following me or just tracing a similar walking route. But if I am suffering from sufficiently degraded executive function, with lower volitional and emotional control and a heightened likelihood of jumping to conclusions, I may wheel around and "defend myself" by preemptively assaulting her.

By the criteria of our theory presented in chapter 3, there is no question that Breivik satisfies criterion 2: he was undeniably the agent behind the carnage in Oslo and Utøya. Brevik's responsibility therefore depends on whether he possessed an MWS at the time of the attacks (criterion 1), and whether his executive processes played the appropriate role in causing them (criterion 3). We are not concerned with the question of whether Breivik was properly diagnosed or even with the question of his case's judicial disposition. What we have tried to show is simply that Breivik's case illustrates the importance of avoiding Moore's "strong-relevance" view, according to which a certain clinical diagnosis would be sufficient by itself to classify Breivik as legally insane, and at the same time resisting the position of Bortolotti et al. (2014) that construes such diagnoses as being of scant, if any, relevance to legal insanity. The key to navigating between these positions, we have argued, is understanding how executive functions enable responsible agency, how those functions can be nonculpably impaired in seriously mentally ill persons, and how those facts help us relate the legal concept of insanity to the medico-scientific concept of mental illness.

The Puzzle of Psychopaths

We now turn to the puzzling topic of psychopaths, who have been an object of public fascination and scholarly interest since the contemporary view of psychopathy emerged in the middle twentieth century. Associated in the popular imagination with figures both real (Ed Gein, Ted Bundy) and fictional (Hannibal Lecter, Anton Chigurh), psychopaths both frighten and mystify us.[8]

Psychopathy is associated with a cluster of symptoms, including manipulativeness, grandiosity, impulsivity, antisocial behavior, and a lack of empathy or remorse. Their inclination to violence, deception, and cruelty tends to be explained not by a generalized inability to distinguish reality from delusion—as might be claimed of a person with schizophrenia—but by an apparent numbness to the features and consequences of actions in virtue of which they are wrong, for example, the suffering of other people. There is evidence, for instance, that psychopaths exhibit poorer ability to detect fear or sadness in facial expressions (Blair 2007). But do these deficits mean that psychopaths are pathologically incapable of competently judging what it would be wrong (or illegal) to do, or do psychopaths make competent judgments that fail to motivate them? Are they "more bad than mad," as the philosopher Heidi Maibom (2008, 176) puts it, or is it the other way around? There would be some strange comfort in feeling that we could neatly classify psychopaths as either "bad" or "mad"—as either basically sane but thoroughly evil persons, or as persons whose minds function in such a way that they cannot be meaningfully considered responsible for their actions. Many feel psychopaths resist such categorization.

A philosophical precursor to contemporary discussions of psychopaths appears in David Hume's example of the "sensible knave"—a *knave* being a common term, in Hume's time, for a dishonest or unscrupulous man. By "sensible knave," Hume means to describe a person who seems cognitively equipped to grasp moral and legal principles but utterly unmoved by the sentiments that encourage our compliance with, and respect for, these norms. The sensible knave claims to understand that one ought to keep promises, for instance, and yet feels no inclination to do so. He may agree that justice and fairness are good things for people most of the time and yet take advantage of the fair dealings of others at every opportunity.

It is important to note that, in Hume's telling, the sensible knave is not uninformed, illogical, or cognitively impaired—that is what makes him *sensible*. For that reason, he can be motivated by prudential concerns. Knavery is likely to get one caught, after all, which hampers the prospects for future knavery. A truly sensible knave, however, will be unmoved by appeal to the welfare or needs of others, and even by appeal to the many forms of human happiness available to him if he will give up his knavery, such as "inward peace of mind, consciousness of integrity ... conversation, society, study, even the common beauties of nature, but above all the peaceful reflection on one's own conduct" (Hume 1975, 283–4). Although the sensible knave might aver publicly that murder is wrong, his "heart [does] not rebel against ... pernicious maxims," and "he feel[s] no reluctance to the thoughts of villainy or baseness" (283).

If Hume's notion applies to psychopaths, as it seems to, then his remarks bear on our inquiry into whether psychopaths are fit bearers of responsibility. Some argue that many psychopaths qualify as legally sane, despite their disordered emotional lives (Maibom 2008). Others argue, however, that psychopaths qualify for (at least) diminished capacity responsibility because they lack the moral knowledge necessary to follow the law.[9] Stephen Morse, for example, claims that a psychopath's lack of emotional data may translate into a lack of legal rationality (Morse 2008b). Psychopathy, by that rationale, should be included in the legal insanity defense as a sufficient mental abnormality to support a claim of legal excuse. We contest this position for two reasons. First, it is not clear that psychopaths really do lack moral knowledge. And more fundamentally, even if psychopaths do systematically lack moral knowledge, that lack by itself does not necessarily deprive them of a fair opportunity to avoid wrongdoing. What *would* undermine the normative competence required for such a fair opportunity is an inability to correct for such a lack via the operations of an MWS. Whether a psychopathic offender has a legitimate claim to an excuse from responsibility, including via a claim of legal insanity, turns on whether or not the offender had an MWS during the time leading up to the crime. It is not clear, however, that psychopaths systematically lack an MWS.

An influential set of studies by Blair (1995, 1997) offered support for a Morse-style position, seeming to show that psychopaths fail to grasp the difference between moral and conventional rules. Interestingly, instead of viewing all transgressions presented as conventional, psychopaths treated both conventional and moral transgressions as not contingent on permissibility, thus indicating they operated like moral, or objective, rules. However, in a more recent series of experiments in which psychopaths were explicitly forced to decide which norm violations were moral and which were conventional, they performed as well as controls (Aharoni, Sinnott-Armstrong, and Kiehl 2012, 2014). And although some earlier studies indicated that psychopaths had reduced prefrontal cortical volume, subsequent findings have complicated this picture, suggesting these prefrontal differences may only be true of some subset of psychopaths: namely, "unsuccessful" psychopaths or those who have regular contact with law enforcement (Yang et al. 2005, Yang et al. 2010). What neurological differences do exist may be less significant for normative competence than was originally supposed: A study by Glenn, Raine, and Schug (2009) found that psychopaths solve moral decision-making tasks by drawing on brain areas different from those recruited by controls, including the dorsolateral prefrontal cortex. Such findings suggest that, rather than lacking moral knowledge, psychopaths may be able to access that knowledge through different means than healthy controls.

A recent review indicates psychopaths do not show abnormal subjective experience of fear, but instead show insensitivity to fear-related cues (Hoppenbrouwers, Bulten, and Brazil 2016). While some psychopaths exhibit a significant deficit in attention, which might account for this insensitivity, there is some evidence that others may be able to appropriately adjust their top-down attention to recognize and respond to affective cues (Larson et al. 2013, Koenigs and Newman 2013). In one study of psychopaths' performance on the gambling task, psychopathic inmates with low attention scores gambled worse than did the attentive ones, whereas in non-psychopathic individuals, attention had no impact (Lösel & Schmucker 2004). Even if psychopaths suffer from flattened affect and a lack of emotional empathy, then, some of them may be able to correct for this deficit via attention.

More broadly, a major problem with using evidence of psychopathy to ground a claim of legal insanity or diminished capacity is that psychopaths appear to be a heterogeneous group with respect to *both* affective and executive impairments. The long-dominant clinical tool for identifying psychopaths, the Hare Psychopathy Checklist (known in its revised form as the PCL-R), does not seem to carve out a group of persons with a common cognitive deficit. This becomes clear when one looks to studies of particular subgroups of psychopaths.

One significant division is between *primary* psychopathy, which is associated with fearlessness, narcissism, low emotional empathy, and social dominance and is inversely associated with negative emotionality, and *secondary* psychopathy, which is associated with negative emotionality, impulsivity, frustration, sensation seeking, substance abuse, and reactive aggression (Skeem et al. 2007, Skeem et al. 2011). Researchers have also drawn a division between *unsuccessful* psychopaths—those who are caught up in the criminal justice system—and *successful* psychopaths—those who are not. Unsuccessful psychopaths—who are, for obvious reasons, easier to study—seem to have reduced prefrontal and amygdala volume (Raine et al. 2000, Yang et al. 2005), reduced autonomic levels (Hare 1965, 1982), impaired fear conditioning (Birbaumer et al. 2005), and normal-to-diminished executive function (Gao and Raine 2010). Successful psychopaths exhibit no reductions in prefrontal or amygdala volume (Yang et al. 2005), show intact or enhanced autonomic levels (Ishikawa et al. 2001), and may actually possess *enhanced* executive function relative to controls (Zeier, Maxwell, and Newman 2009, Zeier et al. 2012).

Thus, primary and unsuccessful psychopaths may suffer from flattened affect, while secondary and successful psychopaths do not. And there is some evidence that unsuccessful psychopaths may have diminished executive functioning, while successful psychopaths may not. However, even within the successful/unsuccessful subsets, the data on specific executive deficits and their contribution to volitional incompetence are mixed. Jurjako and Malatesti (2016) argue that psychopaths may not suffer from deficits in "cool" executive function (the kind tested by "emotion-free" tasks such as the Wisconsin Card Sorting Test), but only in "hot" executive function (the kind probed by emotionally laden tasks such as gambling experiments). And these "hot" deficits, according to Jurjako and Malatesti (2016), can be ameliorated with attentional focus, suggesting that psychopaths with hot executive impairments are not suffering from the kind of global executive incapacity that would support a claim of diminished capacity (Jurjako and Malatesti 2016). Using the "cold" Wisconsin Card Sorting Test, Pera-Guardiola et al. (2016) found that prisoners with lower scores on the PCL-R (indicating weaker prevalence of psychopathic traits) performed worse than healthy controls and those with high PCL-R scores. Thus, higher scores on the PCL-R among unsuccessful psychopaths should not be assumed to indicate greater deficits in cool executive functions.

In the end, we think that the responsibility of psychopaths rests primarily upon the status of their executive functions, and that the PCL-R doesn't tell us enough about the executive functioning of psychopaths to indicate that psychopaths ought to be legally excused, and if so, which ones. We have argued that persons are responsible for breaking the law if they have the capacity to understand moral and legal rules and the volitional capacity to be law-abiding. We believe it is likely, however, that to understand legal rules a person need not feel a certain way about them—a prudential understanding may be enough. A longtime thief who loves stealing and doesn't feel at all badly about breaking the law is still responsible. As Hart (1968) noted, although it is necessary for most citizens to internalize the law for a legal system to be functional, not all citizens will take the law as a good reason to be law-abiding. Those who do not are responsible if they have capacity responsibility or, on our theory, have an MWS.

To put it another way: A color-blind person is not excused from obeying traffic signals simply because he or she lacks normal color perception; if the person is capable of knowing the rules of the road and the relative position of the lights—and volitionally competent to conform his or her behavior according to this knowledge—then the person is responsible for driving within the law. Similarly, a psychopath need not sense the fear or suffering of others, nor experience moral emotions such as guilt or empathy, in order to obey the criminal law—he or she just needs to be able to pay attention to legal rules and which acts violate them, and to conform his or her behavior to them. Until more fine-grained tools can be developed to reliably tell apart psychopaths with a legitimate claim to normative incompetence from those without such a claim, we cannot safely regard a diagnosis of psychopathy as motivating a defense of legal insanity or diminished capacity.

Recall another of the cases with which we opened chapter 1: Brian Dugan, a rapist and serial killer from the Chicagoland suburbs who preyed on children and young women in the 1980s. With their client facing the death penalty, Dugan's attorneys brought in neuroscientist Kent Kiehl to provide expert testimony that might be judged to mitigate their client's guilt. A few years after his involvement in the Dugan case, Kiehl told *Wired* about the differences between psychopaths' brains and those of normal persons:

> We've found that psychopaths have 5 to 10 percent reduced gray matter density in and around the limbic regions [a subcortical network crucial to emotional processing]. We've also found that the tissue that connects the limbic system to the frontal lobes is disrupted. There have also been lots of studies published showing reduced responsivity in those circuits during emotional processing and moral decision making. (quoted in Miller 2014)

Assuming these general findings could be reproduced in Dugan's specific case—according to Kiehl (quoted in Hagerty 2010), Dugan scored in the 99th percentile on the PCL-R, and scans of his brain revealed many of the structural abnormalities characteristic of psychopathy—should such evidence mitigate Dugan's culpability? We suspect not. Recall that, by our theory's criteria, Dugan's responsibility would require that he satisfy criterion 1 (possession of an MWS) and criterion 3 (the appropriate causal role being played by his executive processes); that he caused the harms in question, and therefore satisfies criterion 2, seems unassailable. Even if Dugan indeed evinces flattened affect or impaired emotional processing as part of his psychopathic condition, and even if we could presume he was similarly impaired at the time of his crimes (*decades* before he was subjected to an fMRI), neuropsychological evidence of those impairments *alone* seems insufficient to prove diminished legal capacity. If Dugan's executive processes were functional at the time of his crimes—if he could exercise capacities for attentional control, planning, and inhibition, as some of his crimes suggest he could—then he plausibly had the fair opportunity to avoid wrongdoing by recruiting these executive capacities to control for any affective deficits.

Conclusion

In this chapter, as in the previous one, we have tried to bolster our theory's plausibility by demonstrating its explanatory power. If possessing an MWS is necessary for capacity responsibility (and thus legal agency), then the legal concept of *insanity* should map well onto categories of offenders who lack, because of a mental illness, such a minimal working set. One does not have a fair opportunity to avoid wrongdoing, in Brink and Nelkin's (2013) sense, if one lacks either the cognitive or volitional capacities that make up normative competence. A reasonable interpretation of the insanity defense should therefore include both cognitive and volitional factors, as the MPC standard, but not the M'Naghten rule, does. And the illnesses most commonly associated with successful insanity pleas—schizophrenia and, more rarely, bipolar disorder—are indeed marked by severely degraded executive function, detectable in brain and behavior.

As we see them, executive functions act as a conceptual bridge between the legal concept of *insanity* and the medico-scientific concepts arrayed under the heading *mental illness*. We can thus avoid falling into the dilemma described by Moore (2014): We have to accept neither the implausible view that mental illness is entirely irrelevant to legal insanity nor Moore's strong-relevance position, according to which mental illness, when it excuses someone, can excuse all by itself. Holding to what Moore would call a weak-relevance thesis, we contend that a diagnosis of schizophrenia *can* be relevant to a defendant's plea of legal insanity—not because such a diagnosis is independently exculpatory, but because it may include a clinical judgment of substantial and global executive dysfunction as a symptom of the disease.

Persons who lack an MWS due to mental illness ought to be found not guilty by reason of insanity and be excused from responsibility under the law, according to our approach. Verdicts of "guilty, but mentally ill," which are fairly common in the United States, are inappropriate for those who are legally insane. This is because where a person lacks an MWS due to mental illness, he or she bears no responsibility for their crime, and thus ought not to face punishment. Most legally insane defendants ought to be admitted to a hospital. Without a MWS, persons deemed legally insane are unlikely to be able to manage their own lives, and they are at risk for recidivating.

A diagnosis of psychopathy bears even weaker relevance to a determination of legal insanity than schizophrenia, we argue. Psychopaths compose a heterogeneous class, and although patients with schizophrenia are also a diverse group, schizophrenia is far more closely associated with degraded executive function. The diagnostic tools currently used to study psychopaths are not probing—or are only probing in an indirect way—their executive abilities, meaning that such a diagnosis would not typically include, or even suggest, an assessment of severely compromised executive function. Even very severe psychopaths, as measured by the PCL-R, may have substantially intact executive function. From an evidentiary perspective, a diagnosis of psychopathy by itself would not warrant further psychiatric testing to determine the basis for a legal insanity plea unless there were independent reason to suspect that the defendant had compromised executive capacities.

Psychopathic offenders may exhibit profound deficits in morally relevant emotions such as guilt and empathy, but if these affective deficits are balanced by an MWS, these offenders may be able to conform their behavior to laws and moral norms, and consequently they ought not be excused from responsibility on grounds of legal insanity. The criminal law does not apply to persons who are incapable of understanding or following it, but it does apply to persons who do not care about it or who *cannot* care about it—those whose hearts, as Hume said of the sensible knave, do not rebel against the "pernicious maxims" that might move them to heinous and unlawful acts. The law does not ask the psychopath to *care* about the law—only to obey it.

10 Blame, Desert, and Punishment

There is a clear relationship between blameworthiness and the capacities that are necessary for responsibility. We will argue in this chapter that reasons-responsiveness, understood as baseline executive function, grounds assessments of blameworthiness and basic desert understood as a retributive proportional response to the harm caused. If Sean has capacity responsibility (Hart 1968) and is reasons-responsive (Fischer and Ravizza 1998, Vargas 2013), and he commits an act which results in harm, then Sean is likely to be viewed by the folk as blameworthy for the harm caused, depending upon his intentions when he committed the act. For example, if Sean does not file his taxes, most would feel comfortable claiming Sean's action was blameworthy. *How much* blame is justified, however, would seem to contingent on a host of factors that Hart considered to be a part of liability responsibility, including the specific mental states that a person has regarding the act that caused harm (Hart 1968). For example, if Sean failed to file his taxes in order to avoid having to pay taxes, he is very blameworthy, but if he failed to file his taxes because he forgot about the filing deadline, he deserves less blame. The highest level of blame usually applies to cases where a person actually intended, or desired, the harm caused by his act (in this case, not paying taxes on time). On the other hand, lower levels of blame seem to be warranted where a person knew *there was a chance* harm would result from his or her act or where the person *should have known* harm might result.

Below we will argue that criminal offenders are blameworthy based upon the capacities they possessed at the time harm was caused, and these capacities, in addition to possessing certain mental states causally linked to harm, can serve as the basis for desert (the notion that the person deserves some sort of response to the harm). On our executive theory, for an action to be worthy of praise or blame an actor must possess a minimal working set of executive processes at the time of the act. The amount of praise or blame depends on the appropriateness and level of involvement of the MWS (minimal working set of executive functions) with regard to the action. Intending the harmful result, or having explicit knowledge of the reasonable possibility of a harmful result, increases levels of blameworthiness and amounts to greater involvement of the actor's executive functions. In general then, the greater the involvement of the executive with regard to the harm caused, the more blameworthy an actor is for the harm.

The hierarchy of assessments of blameworthiness mimics the structure of mens rea requirements in the U.S. Model Penal Code, where the most serious crimes such as first-degree homicide require the mental states of "purposely" or "knowingly" causing the criminal harm (American Law Institute 1985). Similarly, under the MPC cognitive disability, diminishment, or incapacity—such as mental illness, intellectual disability, or juvenile status—may serve to downgrade the blameworthiness of an offender, ideally resulting in downgrading of the charged offense (e.g., from first-degree murder to manslaughter) or even rendering an offender exempt from blame and not guilty.

However, the level and type of institutional punishment a criminal offender deserves is a complex matter. The punishment an offender deserves under current law is not just determined by the offender's mental states with regard to the harm, but also by the severity of harm caused: An intentional homicide is punished much more severely than an intentional theft. Further, our current system of punishment is informed by both backward-looking retributive ("just desert") notions of proportionality and utilitarian forward-looking considerations of social order such as deterrence. We will argue that the U.S. model of punishment is best understood as an example of *limiting retributivism*. Under limiting retributivism (Morris 1974), the upper limit of justified punishment is based upon the level of blame an offender deserves for a particular harmful act (taking into consideration the severity of the harm caused). Within this limit, other forward-looking aims may be pursued. We believe our theory of responsibility is compatible with limiting retributivism, and that this type of hybrid justification of criminal punishment can result in just punishments. However, we will claim many U.S. jurisdictions do not adhere to an important principle of limiting retributivism and thus assign criminal punishments that are too harsh.

In the second part of this chapter we will argue that our executive theory offers a unique and powerful critique of the methods of punishment currently used—namely, long terms of incarceration without substantial rehabilitative programming. We will argue that incarceration as it is currently assigned unnecessarily undermines moral and legal agency. We further suggest that this problem is a substantial contributing factor to the high rates of recidivism in the United States. Current criminal sentences offer little opportunity for nonroutine reasoning and decision-making, and thus prisoners' executive functioning—which may already be lower than the average person's—languishes because of underuse. This reduction of moral and legal agency may partly explain prison's criminogenic effect on offenders, and this is a serious worry as the vast majority of prisoners are eventually released. We claim prisoners ought to be given ample opportunity to develop their executive functioning via programs that allow them to learn and practice new skills, such as complex decision making. Prisoners' executive functions also ought to be indirectly supported by offering them opportunities to exercise.

Desert

In this section, we will begin to discuss the way in which our theory interacts with broad justifications for the criminal law and punishment. The criminal law assumes basic desert (Moore 1997, American Law Institute 2007). This means that the criminal law is structured to deliver a deserved response—blame and punishment—to those who commit criminal acts. A basic desert response is backward-looking in that it is justified by features of the actor or the harmful act and not by forward-looking consequences of imposing blame. If our theory that possession of an MWS provides the grounds for criminal responsibility is correct, this means that possession of an MWS also provides grounds for basic desert. That is, on our theory a person must possess the capabilities enabled by an MWS to be fully deserving of moral blame. Here again, we can see the overlap between theories of criminal responsibility and moral responsibility: As we shall see below, many legal scholars argue that the primary justification for the criminal law is to address moral wrongs. One goal of this book is to specify the mental conditions that must be met in order for someone to be morally and criminally blameworthy, and to attempt to understand them in terms of brain processes and functions.

One can find proof of the criminal law's commitment to delivering basic moral desert in the MPC, which was recently revised to state that the primary purpose of criminal punishment is to deliver a retributive response to crime. Retributivism is the notion that criminal punishment is justified if it delivers to an actor his (backward-looking) "just deserts" for the criminal act. The new "purposes" section of the MPC states that the primary purpose of punishment is "to render sentences in all cases within a range of severity proportionate to the gravity of offenses, the harms done to crime victims, and the blameworthiness of offenders."

The principle of retributivism rests upon this notion of proportionality—the idea that punishment ought to be appropriate to the degree of harm caused and the level of culpability for that harm borne by the defendant. Proportional punishment in the criminal law is in part determined by the structure of criminal offenses. Different offenses, which can be committed by defendants with varying degrees of culpability, trigger different ranges of punishments. For example, in the state of Illinois, a defendant who steals an object worth over $500 may be eligible for 1–3 years in prison whereas a defendant who commits a (first-time) sexual assault may be eligible for a sentence of 4–15 years. The range stipulated for each type of crime is the law's first-pass attempt to apply proportional punishment. Proportional punishment is also dependent upon the level of blameworthiness of the defendant. A defendant may be less culpable if he or she has diminished capacity responsibility (or has no capacity responsibility) and is thus eligible for an excuse or justification regarding the criminal harm. A defendant might also be less culpable—or not culpable at all—if he or she is suffering from a severe mental illness. Finally, the blameworthiness of a defendant may also be determined by the tightness of the relationship between the defendant's goals and the crime, which can upgrade or downgrade the severity of the person's crime. For example, a theft might be committed purposely, or just recklessly—for example, a defendant may take a car knowing it belongs to a fellow guest at the bed and breakfast or may take the car with reckless disregard for who owns it (it might be the B&B's car, which the guest is entitled to borrow, or it might belong to a fellow guest). A defendant is more blameworthy for a purposeful theft than a reckless one.

Thus, under a common-law criminal justice system, backward-looking desert involves assessing both the criminal harm caused and the defendant's culpability regarding the harm. The law then aims to deliver a proportional response via punishment. Current common-law legal systems do this via the various structures of the criminal law: the categories of offenses, the gradations of culpability that apply to different offenses, and applicable excuses and justifications. The hierarchy of assessments of blameworthiness in the United States often utilize the categories of mens rea in the MPC (discussed earlier in chapter 4) or some close approximation, where the most serious crimes require that the person "purposely" or "knowingly" caused the criminal harm (American Law Institute 1985). And, as discussed above, the MPC allows that cognitive disability, diminishment, or incapacity—such as mental illness, mental retardation, or juvenile status—may serve to reduce the blameworthiness of an offender, via downgrading the charged offense (e.g., from first-degree murder to manslaughter), or, in rare cases, an excuse may fully exempt an offender from blame and guilt. Finally, at sentencing a judge is often given some discretion to determine the total amount of punishment within a specified range, taking into consideration the factors mentioned above (in addition to others not mentioned, such as prior criminal history).

As indicated above, basic moral desert premises this blame on a person's moral status with regard to the action (by tracking features of the actor and the harmful act) and not on forward-looking consequentialist or contractual considerations (such as future dangerousness or social order) (Pereboom 2013, 168). Basic deserts appear to be grounded in human responsive attitudes to human action such as approval and disapproval (Feinberg 1970). That is, it seems desert is at base an emotional response—a reactive attitude, as discussed in chapter 3—by an observer of a particular actor or act, and blame is the response of disapproval. Punishment, however, is a "derived" desert (Feinberg 1970). Derived deserts are forms of treatment that are fitting expressions of basic deserts: Thus, criminal punishment might be considered a fitting expression of the basic disapproval generated when a defendant commits a serious criminal offense.

Nelkin argues that desert and accountability responsibility—the sort of responsibility whereby we hold each other accountable for actions via blame—mutually entail each other, even though they may appear to involve different demands: "[W]e have, on the one hand, a notion of responsibility (and by implication freedom) understood in terms of appropriate kinds of demands one might make on other people, which seems to be essentially an interpersonal notion, and, on the other, a notion understood in terms of *basic* desert, which seems defined in such a way as to rule out at least one prominent interpersonal account, namely contractualism" (Nelkin 2016, 175). We demand persons obey moral and legal rules when they have the capacity to do so, and it seems we blame them in a backward-looking way when they had the capacity to follow such rules and they fail to do so. Nelkin defines moral desert this way: X's being deserving of sanction is a conditional reason to sanction X. Desert exhibits a particular kind of fittingness, where what is deserved falls on a scale from harmful to beneficial (Nelkin 2016, 179). Accountability has to do with demands that others act in accordance with certain reasons. Nelkin argues the two concepts have the same satisfaction conditions in that, in order to be deserving of praise or blame and to be accountable for an action, an actor must have a *fair opportunity to do the right thing for the right reasons* (Nelkin 2016, 184). The appropriateness of the reactive attitudes that ground desert and accountability require that an agent can meet the demands upon which the reactive attitudes rest, which she and David Brink argue involve fair opportunity abilities (Brink and Nelkin 2013). Fair opportunity is scalar: "[T]he less fair an opportunity and the less serious an action or omission, the less blameworthy in the accountability sense, and the lesser the sanction deserved" (Nelkin 2016, 185).

We agree that being deserving of sanction depends on the capacities that allow us to be moral rule followers and law abiders, and that such capacities are scalar. Specifically, in chapter 4, we argued that to be deserving of a criminal sanction (1) the agent must have had an MWS; (2) the agent must have caused the criminal harm; and (3) the agent's MWS must have played an appropriate role in the genesis of the action, omission, or consequence causally related to the criminal harm. Typically, an agent is deserving of sanction for an action where the action was causally related to the synchronic operations of the agent's MWS which we referred to as "executive engagement" in previous chapters. More severe sanctions are deserved where the criminal harm was explicitly represented by the agent, or was an explicit target of the action, than in cases where the criminal harm was not contemplated by the defendant. As we discussed in chapter 4, we agree with Nelkin and Brink although we understand "fair opportunities" in accordance with our executive theory: If an actor has an MWS, and thus the capacity for reasoning, planning, and inhibition, the actor had an opportunity to avoid the criminal harm (or to pursue it).

In general, the greater involvement of the executive in the action results in higher levels of accountability and desert for an action, because in this case the person either pursued the harm with his or her action or at least was aware of the relationship between the action and the criminal harm and thereby had greater top-down control over the action that caused the harm. As Levy notes, when an actor is aware of the moral implications of his or her act, he or she has a robust opportunity to pursue or inhibit the act based upon this knowledge (Levy 2014). On the executive view, in cases where an offender acts "purposely" with regard to the criminal harm, or "knowingly" (knowing that criminal harm would result), the offender's executive functions directly interact with information about that harm. In this case an agent applies top-down attention to the possibility of criminal harm and has the opportunity to make plans with regard to that harm, recall memories related to the harm, and manage emotional input regarding the harm. Then the agent may inhibit or initiate the action.

An actor who is less culpable and deserving of sanction may have engaged his or her MWS with regard to the criminal harm only in a very shallow way, or could have subjected the possibility of criminal harm to his or her executive processes in a counterfactual sense but failed to do so (e.g., *the person should have known* that the criminal harm would result from the action; the person *would have*, had the person paused and thought the action through). In both cases the agent has less opportunity to utilize his or her executive processes such that he or she would have acted in a law-abiding way. We discussed such negligence cases in some detail in chapter 7, where we contrasted our handling of this type of case with Neil Levy's. Holding an offender responsible for committing an act that causes criminal harm *negligently* generally requires that the agent possessed an MWS in the period running up to the action, such that the diachronic operations of the offender's MWS could have resulted in his or her paying attention to the possibility of criminal harm, and the person could thereby have inhibited the action or acted to prevent the consequence and avoided the criminal harm. Where a defendant shows "reckless disregard" for the criminal harm, the defendant's executive processes may have dismissed the possibility of the harm when planning the action. In such cases agents did not have as robust of an opportunity to avoid the criminal harm as offenders who act for the purpose of causing criminal harm, but they still deserve some blame for violating the law when they had opportunity not to do so, given their cognitive capacities to understand moral and legal rules and to abide by them. Where a defendant is negligent, a reasonable person would have known that a criminal harm was a risk brought about by the behavior, and the agent should have organized his or her behavior such that the agent would have engaged his or her attentional capacities to recognize the risk.

Desert Skeptics

Most contemporary legal scholars and philosophers believe that the law's assumption of basic desert is compatible with current knowledge regarding the psychological states and processes possessed by persons who commit crimes.[1] As we noted in chapter 3, even if such states and processes are in some way determined (at the level of neurons and other cells or at the level of actions), we may still say that persons who commit moral or immoral acts—or certain legal or illegal acts—deserve praise or blame. Many contemporary philosophers are compatibilists who believe determinism is irrelevant to what is really important about free will. Compatibilists believe that the attribution of desert for actions can be justified if an agent possesses certain cognitive capacities, which are necessary for responsibility (Nahmias 2014, 6). We agree: Possession of an MWS is necessary, but not sufficient, for attribution of responsibility, including criminal responsibility.

In fact, we would go one step further: There is no need to appeal to free will at all to justify basic desert and holding a person responsible. Much of the debate surrounding free will and responsibility concerns defining free will as something that either is, or is not, possible for humans to have. However, we have argued here that what matters is not the conditions or capacity necessary for free will, but responsibility directly. Attempts to prove the existence of free will have been distracting philosophers of action from what really matters: establishing the exact grounds for praise and blame. We embrace this project and don't feel the need to claim a particular position regarding the existence of free will. Human beings may or may not be free, depending upon the definition of freedom one uses; even so, they are deserving of praise and blame, and in some cases punishment, depending on whether they possess certain cognitive capacities and psychological states.

It would be wrong to dismiss the views of the handful of philosophers who are desert (and thus responsibility) skeptics too quickly, however. It is possible that their arguments show that the cognitive capacities we feel ground responsibility either do not exist or are not sufficient to ground responsibility. Thus, we will take some space here to explore in a bit of detail the position of contemporary desert skepticism.

Desert skeptics argue that desert requires a fact be true about agents—a fact concerning their freedom of will—that is in actuality radically false. Thus, they claim criminal justice systems are not justified in punishing offenders based upon moral desert. According to a recent version of desert/responsibility skepticism, held by Derk Pereboom and Gregg Caruso, the problem with free will is not related to determinism, despite the popularity of challenges to free will based upon determinism. Instead, Pereboom and Caruso (2018) claim persons do not have free will regardless of whether the universe turns out to be largely deterministic or somewhat indeterministic. They argue compatibilism and libertarianism about free will are both false because neither provides an account of agential control sufficient to vindicate the truth of free will. They thus advocate for the position of hard incompatibilism:

> Against the view that free will is compatible with the causal determination of our actions by natural factors beyond our control, we argue that there is no relevant difference between this prospect and our actions being causally determined by manipulators (see Pereboom 2001, 2014a). Against event causal libertarianism, we advance the disappearing agent objection, according to which agents are left unable to settle whether a decision occurs and hence cannot have the control required for moral responsibility (Pereboom 2001, 2014a; Caruso 2012). The same problem, we contend, arises for non-causal libertarian accounts, which also fail to provide agents with the control in action required for basic desert moral responsibility. While agent-causal libertarianism could, in theory, supply this sort of control, we argue that it cannot be reconciled with our best physical theories (Pereboom 2001, 2014a) and faces additional problems accounting for mental causation (Caruso 2012). Since this exhausts the options for views on which we have the sort of free will at issue, we conclude that free will skepticism is the only remaining position (199).

Because Pereboom and Caruso argue that an action or attitude cannot be properly assigned to a particular agent as his or her own decision or attitude (for which he or she is responsible) under a determinist or indeterminist worldview, no agent ever deserves backward-looking praise or blame (or punishment). This means basic desert is unfounded, and so is retributive punishment. As discussed above, accountability responsibility is grounded in backward-looking reactive attitudes such as resentment, blame, or agential anger. However, the folk concepts of blame and praise aim at a fictional entity (agential control) and thus fail to refer to anything real, like the concept "witch," according to Pereboom and Caruso (2018). If the state wants to address worries about an agent's potential to cause harm or break laws in the future, they claim, it must do so by appealing to purely forward-looking justifications such as incapacitation or rehabilitation.

We will focus here on the claim by Pereboom and Caruso that agents cannot exercise meaningful control over their actions such that they warrant praise and blame for them. We grant Pereboom and Caruso that basic desert depends upon the existence of agent control, and thus if such control does not exist, basic desert ought to be eliminated. We claim the human capacities enabled by an MWS give rise to meaningful diachronic and synchronic agential control, and it is this control (among other capacities) that justifies basic desert. Given our focus on criminal responsibility, it is easiest to see the work that possession of an MWS does in cases of responsibility based upon negligence because negligence cases do not require a synchronic desiring for, or knowledge of, the criminal harm caused. In this way, we can see how diachronic possession of an MWS alone can ground basic desert.

In chapter 7 we discussed a negligence case involving Bert, the father who forgot he had custody of his kids one weekend and flew to Las Vegas instead of watching them. The kids, ages six and eight, spent the weekend alone in his apartment, and as a result Bert was arrested for child neglect. In that chapter we argued that agents can exercise top-down diachronic agential control over mental processes and the way in which they generate action. In this way, diachronic self-authorship can ground a compatibilist account of responsibility that preserves real agency but avoids "obscure metaphysics" (Roskies 2012, 323). We agreed with Roskies's claim that we, as human beings, can deliberately intervene in our future selves and thereby be responsible for who we are and our behavior (Roskies 2012). Bert might have avoided his lapse by utilizing a hard rule of writing down every aspect of the schedule with his children, or by engineering his environment such that he was less likely to forget his parenting schedule (set a reminder or kept a calendar), or by practicing deliberative self-control by slowing down and mentally reviewing his schedule before making decisions that might impact his parenting schedule. Because Bert had capacity responsibility, and thus an MWS, before his lapse, his executive processes should have performed manipulations on the way in which subconscious mental processes generated his action.

Pereboom and Caruso claim that nothing is ever within agents' control such that they are deserving of praise and blame. We think the above case of Bert provides a robust counterexample. Persons do indeed exercise diachronic agential control, and when we look a bit more closely, capacity responsibility rests upon both diachronic and synchronic cognitive abilities—grounded in executive functions—allowing persons to understand and behave in accordance with legal and moral rules. Capacity responsibility, in turn, grounds basic desert. In negligence cases, legal liability responsibility requires diachronic capacity responsibility (grounded in the operations of an MWS over time) plus violation of a legal standard of care, and in most other cases, legal liability responsibility requires capacity responsibility (grounded in the operations of an MWS over time) plus a specific mental state with regard to the criminal harm, operated upon by executive functions, related to a voluntary act.

We know Pereboom and Caruso are unlikely to be persuaded that the sort of agential control discussed above is sufficient to ground basic desert. Bert's ability to perform diachronic interventions is no doubt governed by his genetic predispositions to possess a certain executive profile along with many environmental factors (past and situational). Bert's genes, his past and present environment, and the cognitive capacities built based upon their interactions have shaped Bert's ability to use his MWS to perform moral actions and be law-abiding. As with epistemological skepticism, at some point it may make sense to wave one's hands in the direction of the free will skeptics and admit that one has not proven what some skeptics wanted us to prove: If freely willed actions require an agent to be firewalled off from his or her causal influencers when acting, or requires that a causal chain somehow originate out of nothing in the agent, we have not shown how this can be the case, and indeed, we do not believe it can be shown. Nothing—no causal powers available to persons within our universe—could satisfy the free will skeptics. But even so, we feel it is fair and important to blame Bert and other agents with an MWS because their executive processes enable them to have control over their actions over time, in direct and indirect ways, such that they may understand and follow moral and legal rules.

Justifications of Punishment

It is now time to turn our focus away from blame and desert to institutionalized punishment (or derived deserts). Criminal sanctions, including incarceration, are designed to serve particular functions. These are often called the principles of punishment, and the four primary functions are retribution, deterrence, incapacitation, and rehabilitation. Above we focused upon moral blame and desert, which serve as the basis for the principle of retribution. The principle of retribution aims to give violators of the law their "just deserts" such that punishment provides a harmful response to a wrongful act. The principle of deterrence, on the other hand, is forward-looking and attempts to influence an offender's decision-making with the threat of punishment. Both the general population and the specific offender who is punished may be deterred from choosing to commit criminal acts by punishment. The principles of incapacitation and rehabilitation are also forward-looking. Incapacitation also aims to stop defendants from offending, but there is no attempt to influence decision-making; instead the offender's environment is manipulated to make reoffending impossible, typically via incarceration. Rehabilitation is the practice of attempting to reform offenders so that they will not reoffend.

Many legal scholars agree that criminal punishment aims to fulfill these multiple functions although adherents of different broad ethical theories emphasize the importance of different functions. Deontological moralism claims the law is justified as a means to address moral wrongs and thus stresses the aim of retribution, where punishment is based upon blame and must be proportional to the moral wrongfulness of a crime. On this view criminal offenders deserve moral condemnation and punishment proportional to the harm caused by, or moral wrongfulness of, their actions. On the other hand, some—including free will skeptics Pereboom and Caruso—argue the law is primarily justified as a means to achieve social order, and the forward-looking aims of deterrence, incapacitation, and rehabilitation are easily understood as supporting this sort of utilitarian justification.

As Brown notes, one of the central problems in the criminal law is that it does not seem justifiable using a single ethical theory (Brown 2002). Western systems of criminal justice tend to embrace all four of the functions listed above, and the relationship between the different aims is complex. In the United States, the recent reworking of the MPC's justifying aims reflects a renewed commitment to retribution as the primary principle of punishment. We support this focus on retribution because we hypothesize that moral desert is grounded in natural human responsive attitudes to human action (Feinberg 1970) and, like Brink and Nelkin, we feel that blame correctly tracks important features of human agency. Thus, we believe that an assessment of moral blameworthiness ought to act as a general constraint on the total amount of punishment that can be applied in any case. However, we feel the other principles, especially incapacitation and rehabilitation, are also important and may inform the type of punishment that is applied within the range of appropriate punishment proportional to the offender and his or her wrongdoing. That is, the total amount of punishment must be proportionate to the crime and to the type of offender (e.g., homicide vs. theft, adult vs. youth offender, offender with full mental capacity vs. diminished capacity); however, questions regarding whether the offender needs to be incapacitated via incarceration, or whether certain types of punishment are likely to deter other offenders similar to this offender in the future, or whether certain punishments will rehabilitate (or will reduce the possibility of recidivism) should also be considered within the overall parameters of proportionality.

Norval Morris famously advocates this kind of account (Morris 1974), often called limiting retributivism, and some argue that it is the consensus model of criminal punishment in the United States and Europe (Frase 2004). Limiting retributivism is a hybrid theory of punishment, where retributive notions of just desert provide an appropriate range of justified penalty within which an offender might be sentenced. Backward-looking retributive considerations of proportionality must then be balanced with forward-looking considerations of social order to create a punishment package that is proportional to crime and offender but also aims to reduce recidivism and overall crime rates.

Morris's limiting retributivism specifically places strict upper limits on punishment based on desert, but no lower limit (Frase 2004). Morris also endorses the principle of parsimony in punishment, which requires that "the least restrictive sanction necessary to achieve defined social purposes should be imposed" (Morris 1974, 59). Courts ought not to impose the maximum an offender deserves unless there are very good reasons to do so and, indeed, should aim to assign lesser sentences and community-based sanctions whenever appropriate (Frase 2004). Reasons to impose a sentence toward the more severe end of the retributive range include forward-looking considerations such as the need to incapacitate an offender considered especially dangerous.

Many important legal scholars have adopted some version of a hybrid theory of criminal punishment, including H.L.A. Hart, who also viewed desert as providing an upper limit on criminal sanctions (Hart 1968, 237). Hart indicated that one must appeal to a retributive account of appropriateness of punishment given the crime committed, which "set[s] a maximum within which penalties, judged most likely to prevent the repetition of the crime by the offender or others, are to be chosen" (Hart 1968, 236–7).

There are worries regarding how any retributive theory, including limiting retributivism, can distinguish wrongful behavior deserving punishment from behavior which the state ought not punish (e.g., adultery) and how such a theory can clearly articulate degrees of blame and wrongfulness. However, we agree with Frase that the criminal law, especially U.S. state law, already does an acceptable job of providing a proportional structure of offenses (Frase 2004). There is general agreement that state criminal codes address behavior that ought to be considered criminal and deserving of criminal punishment, and do so utilizing a sliding scale matching blameworthiness and wrongfulness of an act to degree of punishment. As discussed above, one way a legal system measures blameworthiness is by determining the closeness of the relationship between the criminal harm caused and the offender's intentions. For example, if an offender acted for the purpose of causing the criminal harm, that offender is deserving of more blame (and thus punishment) than if the criminal harm was an unforeseen result of an act. We also noted that on our theory, this means that if an offender's executive processes specifically engaged with information about the possible criminal harm, then the action is more blameworthy because the offender had an explicit understanding of the harm and could have inhibited the action, or engaged in another plan of action that would not have resulted in harm, and so on.

Another way the law determines what amount and type of punishment are warranted is by assessing aggravating factors. Whereas the law's judgment of the relationship between the offender's mental states and the criminal harm is largely an attempt to examine backward-looking proportionality, aggravating factors may be used to assess both an offender's blameworthiness and his or her future dangerousness. The Illinois Criminal Sexual Assault Act (720 ILCS 5/12–12, et seq.), for example, stipulates a wide range of possible sentences for conviction of a sex offense (defined as sexual penetration with force or threat of force) from four years to natural life, depending on the presence of aggravating circumstances. Such circumstances include whether this is a defendant's first offense, whether the victim suffered bodily harm, whether the offender used a deadly weapon, and the age of the victim (both a minor and an elderly victim enhance the sentence). Some of these make the action more wrongful; others, such as information about the offender's criminal history, may indicate that the offender is more likely to be dangerous in the future. Thus, under the statute a first conviction of simple sexual assault might result in anything from a 4-year to a 16-year prison sentence, and aggravating factors may further increase the length of sentence: Sexual assault resulting in bodily harm or of a young victim carries a sentence of 6 to 30 years on a first conviction, or natural life if the offender is being sentenced for a second sexual assault.

Our theory grounds some of the most important considerations with regard to proportional punishment within the brain, and it claims introduction of evidence regarding an offender's MWS and executive function may be important to determining the amount of punishment he or she deserves. Above we discussed the way in which the degree to which a defendant's executive is engaged with information about the possible criminal harm might indicate higher levels of blameworthiness. In previous chapters we also discussed specific cases of excuse. In chapter 8 we argued that due to their lack of an MWS (at very young ages) and diminished executive capacity (in their teens) juvenile offenders ought not to be punished as adults. This means that juveniles deserve less and different types of punishment in comparison to adult offenders. From a retributive perspective, juveniles are less blameworthy, and given the forward-looking aims of punishment, they ought to be punished in a way that is likely to assist in the robust and healthy development of their MWS so they have the capacity to be law-abiding as adults.

Similarly, in chapter 9 we argued that offenders with a mental illness that resulted in their lacking an MWS at the time of their crime are less or even not at all blameworthy for their crime. We believe that offenders who lack an MWS at the time of their crime because of a mental illness ought not to be found "guilty but mentally ill" as is possible by statute in many American states because their lack of mental capacity means they are not culpable for their act and thus no amount of criminal punishment is proportional. Such offenders may only be subject to admittance to a hospital for the mentally ill as would be justified on a civil model of involuntary treatment.

Finally, our discussion of psychopathy in chapter 9 indicated that classification of an offender as a psychopath using the current diagnostic, the PCL-R, does not tell us enough about whether that offender had a functioning MWS at the time of his or her crime to be relevant to determinations of criminal culpability. Although some psychopaths suffer from affective deficits, such deficits may not impact capacity responsibility unless they also lack the ability to correct for that deficit via their executive functions. Fine-grained tests of executive function might be used to assess offenders suspected to be psychopaths as a means to determine whether they possess full capacity responsibility and are deserving of full responsibility and punishment for their crimes.

Amount of Punishment Deserved

Our resources are misspent, our punishments too severe, our sentences too long.
—U.S. Supreme Court Justice Anthony Kennedy

Above we have argued that our theory may ground a limiting retributivism model of criminal punishment, where the total amount of punishment is dependent upon factors of blameworthiness. Such factors include not only the type of harm caused (e.g., a theft vs. a rape), but also whether the offender had full mental capacity (i.e., possessed an MWS) at the time of the crime; whether the offender intended or knew the criminal harm would result from his or her act (i.e., whether there was synchronic executive engagement of the right type); and whether the offender could have avoided the criminal harm (via diachronic use of his or her MWS). On a limiting retributive model, once a fair maximum amount of punishment has been established via such backward-looking factors, an offender's punishment may then be informed by forward-looking considerations to either decrease the total amount or inform what type of punishment is appropriate. Below we discuss the way in which different methods of punishment might serve or undermine these forward-looking aims via their impact on an offender's MWS.

However, before moving on to our discussion of specific methods of punishment, we want to note that our model supports substantially shorter sentences across all categories of crime than are currently standard in the United States. American prison sentences are notoriously harsh. Although the United States only has 5% of the world's population, it houses 25% of the world's prison population (Lee 2015). The United States has the highest prison population rate in the world at 716 per 100,000 people. More than half the 222 countries and territories in the World Prison Population List, compiled by the U.K.-based International Center for Prison Studies, had rates below 150 per 100,000 (Walmsley 2015). Importantly, this is the case even though general victimization rates in the United States rank about the same as those in countries in Western Europe. The United States ranked about midpoint in homicide rates in 2004 to 2012, compared to more than 100 countries, but still had the highest incarceration rate in 2010 to 2013. In addition, nearly half of federal prisoners incarcerated on September 30, 2015, had been sentenced on drug crimes (Carson and Anderson 2016).

Persons found guilty of a crime serve substantially longer sentences in the United States than in many other first-world and Western countries. So-called tough-on-crime policies were enacted in the 1980s and 1990s, resulting in longer sentences, including truth-in-sentencing laws, mandatory minimums, mandatory drug sentences, life sentences without possibility of parole, and three-strikes laws, which often require mandatory life sentence for a third felony conviction. Life sentences have become so commonplace in the United States that about one out of eleven people imprisoned is serving one, and nearly one-third of those are serving a life in prison without the possibility of parole, or LWOP, sentence (de la Vega et al. 2012). The United States is in the minority of countries using several sentencing practices, including life in prison without parole, consecutive sentences, juvenile life without parole, juvenile transfer to adult courts, and successive prosecution of the same defendant by the state and federal governments.

Although the American penal system generally adheres to the limiting retributivist model, many of the sentencing practices above fly in the face of Norval Morris's most important requirement: the limiting principle of humaneness and economy which he called parsimony (Frase 2004, 2009, Morris 1974). This principle was meant to ensure that punishments be no more severe than necessary to achieve the aims of punishment. "In practice, this means that judges should start at the low end of the range of deserved penalties and only increase the sentence as needed to achieve non-desert-based sentencing purposes" (Frase 2009, 255). The use of mandatory minimums and drug sentences, three-strikes rules, and the overuse of life sentences have resulted in sentences that are not proportional to blameworthiness and violate the principle of parsimony. Further, in practice, forward-looking considerations often do not inform prison sentences in a rational way. The United States relies very heavily on jails and prisons and much more rarely utilizes community- or social service-based sentences. This means some offenders who otherwise would have been diverted to noninstitutionalized care are instead sent to prison. One reason may be that in the United States there are often not good local options for social services and rehabilitative programs even for the general population, much less for those convicted of a crime. Due to a lack of alternative sentencing options, for many sentencing courts incarceration is the only option. Where available, mental health services, addiction interventions, or job training may open up the possibility of a shorter sentence of incarceration via rehabilitation of the offender (and decreasing his or her chance of recidivism); where they are not available, a long prison sentence seems the only option. And in cases of mandatory minimum sentences, the court has little discretion to consider forward-looking aims.

Additionally, it seems that incarceration may have a criminogenic effect, especially longer sentences. Several recent scientific studies have focused on the effects of imprisonment, and their findings indicate that prisons do not reduce recidivism more than noncustodial sanctions. Across all offenders, prisons do not have a specific deterrent effect, and low-risk offenders are most likely to experience increased recidivism due to incarceration (Cullen, Jonson, and Nagin 2011). Reviewing some of the current research, Cullen and colleagues conclude that "[t]o the extent that prisons are used because of the belief that they reduce reoffending more than other penalty options ... this policy is unjustified" (Cullen et al. 2011, 51).

We feel that the American criminal justice system generates sentences that often violate the principle of parsimony and thus are inappropriate under a limiting retributivism model. In addition, the overly harsh sentences may actually undermine the forward-looking aims of punishment by having a criminogenic affect. Below we explore the way in which long sentences of incarceration may impact moral and legal agency given our executive theory, and we discuss some of the ways in which community-based and rehabilitative sentencing might enhance the executive capacities underpinning such agency. Our theory, we think, provides some new ways to think about why prison sentences in the United States tend to have a criminogenic effect, and it can shed some light on how sentences might be altered to decrease this effect.

"Prison Brain" and Methods of Punishment

Across the world, recidivism rates among released prisoners range between roughly 30% and 65%, depending on the country where a prisoner serves his or her sentence (Meijers et al. 2015). In the United States, nearly half of federal prisoners were rearrested within eight years of their release, and a third are reconvicted (Hunt and Dumville 2016). Some factors known to contribute to recidivism include age (young) and sex (male), underemployment, substance abuse, and prior convictions (Meijers et al. 2015). Recently there has been a proliferation of research into the cognitive factors that might increase chances of recidivism, with a focus on executive functioning. This literature recognizes that executive functions are crucial for top-down decision-making and self-regulation and thus are necessary if persons are to recognize and abide by moral and legal rules (Meijers et al. 2015, Hofmann, Schmeichel, and Baddeley 2012). A 2015 review article on so-called "prison brain" states that "[u]pon re-entry in society, prisoners face many challenges that place a demand on executive functions.... Consequently, executive dysfunction may cause an increase in reoffending, through failure of self-regulation" (Meijers et al. 2015, 1).

The review by Meijers et al. found many distinct executive function deficits in the prison population, including deficits in "attention, set-shifting, working memory, problem-solving, and inhibition" (Meijers et al. 2015, 3). The way in which diminished executive functioning feeds into recidivism is complex, however. Persons sentenced to prison may have diminished executive functioning at the time they commit their crime, and thus their diminished functioning might have been a factor that led to their first prison sentence. In other words, it is possible that some persons currently deemed to have full capacity by the courts and sentenced to prison in fact have diminished executive functioning. A small set of these offenders probably should have been deemed to have diminished mental capacity by the courts because, on our theory, they do not meet the minimum standard of functioning we have identified as an MWS. A much larger set of these offenders, however, probably do possess an MWS and thus are fully culpable, but still have less executive capacity than the average person. On our view, the former, smaller set of offenders without an MWS ought to be punished less harshly than offenders with full legal capacity (who do have an MWS). Offenders with some executive deficits but with an MWS, we claim, are fully culpable and thus deserve a sentence in proportion to their crime.

In either case, where offenders enter prison with diminished executive functioning, this diminished function may be a factor leading to recidivism once an offender is released from prison. The prison environment may then have the effect of further diminishing a prisoner's executive functioning. First, executive functioning is strongly correlated with physical activity: Increased physical activity improves executive functioning across all age groups (Hillman, Erickson, and Kramer 2008). Yet prison is characterized by physical inactivity. Prisoners spend most of their time in passive leisure activities, such as watching TV, or sleeping (Elger 2009). One study found prisoners spent over nine hours sitting or lying on their beds besides the time they spent sleeping (Ireland and Culpin 2006). There is some evidence that a lack of physical activity may have a negative effect on the prefrontal cortex, and ample evidence that regular engagement in exercise can provide a simple means for healthy people to optimize a range of executive functions (Guiney and Machado 2013). Thus, the lack of opportunity for physical activity in prison might result in diminished executive functioning in inmates. This may be one reason why studies show yoga improves executive functioning in inmates, in addition to the possibility that yoga can directly diminish anxiety and improve attention, memory, and behavioral control (Bilderbeck et al. 2013).

Second, long sentences of incarceration may directly impact the executive functioning of offenders because the impoverished environment of prison leads to diminished use of executive capacities. Executive functioning is positively related to environmental enrichment (Meijers et al. 2015). However, prison life is characterized by a lack of control and decision-making by inmates: what to eat, what to do, where to go, and who to see are all daily activities that are largely decided for offenders. Incarcerated offenders have little opportunity to exercise the cognitive skills of attentional focus, problem solving, and appropriate task switching. Just as many other cognitive skills can become diminished in accordance with the dictate "Use it or lose it"—fluency in a language, the ability to play an instrument, and the ability to solve certain types of problems—so too might the functioning of broader cognitive capacities suffer from disuse. Inmates released from prison often describe feeling overwhelmed by the amount of important decisions that have to be made once they are released: After having made very few decisions of consequence for years, it may be especially difficult for them to go through the cognitive process of determining the best action in a complex real-world situation.

Aristotelian virtue theory offers a more nuanced way to understand how long periods of incarceration may impact executive functioning, and it may lend support to the notion that executive functions may become diminished with disuse. Although we cannot defend this claim in any detail here, we think executive functions are vitally important to Aristotle's notion of practical reason. Aristotelian virtue theory claims that character traits, which can be thought of as stable dispositions to act, develop from habituation. Habituation involves practicing the trait via the use of practical reason, which allows a person to determine which actions are appropriate in a given situation. A disposition to act in accordance with a trait such as honesty is thus ingrained as a result of making appropriately honest choices over time. Creating a stable character trait, such as honesty, is similar to developing expertise. This is a point Aristotle makes explicitly in the *Nicomachean Ethics*, and which Julia Annas (2011) elucidates in her book *Intelligent Virtue*. An expert electrician will know how to solve novel problems and will be able to articulate to a novice why a particular solution is the right one (Annas 2011, 19). Similarly, expertise in a character trait such as honesty will mean that a person can apply the trait in new or difficult situations and later provide reasons as to why he or she acted in such-and-such way.

The executive functions that make up an MWS—planning, memory, attention, inhibition, and task switching—are crucial to picking an act in support of a desired character trait in novel or varying situations. Virtue theory claims that, practical reason, and, we think, executive functioning, is therefore critical to the process of becoming a virtuous or vicious agent. One "becomes just by doing just actions, temperate by doing temperate actions, brave by doing brave actions" (Aristotle 1985, 1103b). Executive functions are thus also important to establishing dispositions to act, especially for supervising the mental processes needed to begin the habituation process. From this perspective we can see that the long prison sentences handed out in the United States, especially LWOP sentences, may undermine prisoners' moral agency by diminishing inmates' ability to develop their character in two ways. First, prison denies inmates meaningful choices—they have very little opportunity to pick one action over other viable options as a means to establish a character trait. For example, if inmate Saul is not given the opportunity to make a meaningful promise or to keep important commitments, he thereby lacks the opportunity to establish the trait of being trustworthy. He also cannot choose to be diligent in doing a job well and thereby lacks the opportunity to establish the trait of being hardworking. Indeed, very long sentences may deny inmates a range of situations and choices over long periods of time such that any already established character traits become undermined. For example, Saul may have had prior well-established dispositions to act both virtuous and vicious—let's say, the trait of being a good father and the trait of being bad with money—both of which may become unseated during a long stint in prison. And note, even though it may be a good thing that his vicious trait of being bad with money may become unstable (the disposition becomes less established), traditional incarceration is not helping Saul learn to be better with money and thereby become more virtuous.

Further, the lack of meaningful choice in prison may not just negatively impact inmates' character or dispositions to act, but also affect their practical reasoning capacities broadly speaking. Practical reason and executive functions may be thought of as cognitive skills that can be honed with practice. For example, planning is certainly a skill that can be improved upon over time with practice; so too is managing one's emotions and paying attention to the right things or switching attention to something new when necessary. As parents know well, children can learn to slow down and better imagine the consequences of an action, to calm themselves down before acting, and to pay better attention to their parent/teacher/peer.

Adults may also learn to improve their organizational skills, participate in anger management therapy, or make better decisions via cognitive behavioral therapy, or even meditation. In contrast, such opportunities are rarely available to prisoners, and very few situations arise in a prisoner's life such that he or she is asked to use the tools of practical reason. As a result, such tools may become dulled.

These arguments apply even more convincingly to sentences that involve long terms of solitary confinement. Whereas all prisons are an impoverished environment, isolation sanctions can almost completely remove an offender's capacity for choice making or any sort of practical thinking. All versions of solitary confinement mean living 23 to 24 hours a day in a cell. Some inmates are granted one hour for exercise, which usually takes place alone in an exercise room or a "dog run" (Haney 2003). Supermax prisons, in particular, are specially designed facilities for isolation and can be either stand-alone or connected to larger prisons. Supermax prisoners are confined to single cells around the clock, released only three to five hours a week for showers or exercise, and subject to constant surveillance (Lovell, Johnson, and Cain 2007).

There is evidence that prisoners in solitary confinement at supermax prisons fare even worse than prisoners in the general population once released. A 2007 study found that prisoners released directly from a supermax prison committed new crimes sooner than prisoners who were transferred from segregation to the general population for several months before being released (Lovell, Johnson, and Cain 2007). Some prisoners within the general population retain choices regarding who to talk to, what to read or watch, and whether to exercise and follow the rules. Such prisoners may take advantage of these little freedoms to work to develop virtuous character traits, or at least continue to make some choices regarding their interactions with other prisoners. However, in the case of isolation sanctions, even these small opportunities are often removed. The forced idleness of isolation can deny offenders any choices short of the position they sit, stand, or lay in and whether to eat the food delivered to them (a hunger strike is one of the few ways they have to express dissatisfaction). This complete lack of choice doesn't just halt the use of practical reason and character development, but can lead to serious psychological hardship. As one offender subjected to the practice noted: "I went to a standstill psychologically once—lapse of memory. I didn't talk for 15 days. I couldn't hear clearly. You can't see—you're blind—block everything out—disoriented, awareness is very bad … I think I was drooling—a complete standstill. I never recovered" (Guenther 2012).

Better Methods of Criminal Punishment

We think methods of criminal punishment used under a limiting retributive model ought to aim to enhance, or at least not diminish, an offender's means to moral and legal agency wherever possible. This means community-based or alternative sanctions ought to be used in many cases where an offender is convicted of a nonviolent crime. We also think that jails and prisons (but especially prisons) ought to try to provide opportunities for inmates to be physically active and to make meaningful choices such that they can use their practical reasoning and executive functions. Although under a retributive model a person punished is meant to suffer some harm, the sort of harm is not determined by the model. Western nations have landed upon incarceration as the primary harm imposed upon offenders for several reasons; denial of certain liberties has become the primary harm to be delivered, and corporal punishment has been rejected. Denial of liberties is a sensible means of criminal punishment, and both community-based sanctions and, in some cases, incarceration are acceptable ways to deny liberties. However, both may be designed to also serve forward-looking aims of punishment.

Nonprison, community-based sanctions for suitable offenders are much less costly than prison and reserve prison space for the most dangerous offenders. Importantly, community-based sanctions allow prisoners to stay in their homes and maintain control over large portions of their lives: Offenders serving community-based sentences can make their own daily decisions, earn a living, and maintain social contacts and networks, which means that they may continue to exercise their moral and legal agency in meaningful ways. Intermediate supervision options such as electronic monitoring (where an offender may live at home but have curfews enforced via electronic ankle monitors), residential programs (where offenders may live in a residential home for offenders, but pay rent and may hold a job), and specialized problem-solving courts (such as drug courts) are less costly than incarceration and provide a greater degree of monitoring and requirements than traditional probation or parole programs (National Conference of State Legislators 2011). Special court systems are often particularly effective:

They attempt to provide extra supervision and programming aimed at an offender's particular risk factor for recidivism (youth, a drug addiction, or a mental disorder). These special factors often make an offender more deserving of community-based sanction in comparison to other offenders because his or her special status (as a young, drug addicted, or mentally ill offender) is often partially exculpatory. Regular prison will often exacerbate the problems of such special offenders; in contrast, if an offender's needs are addressed within the community, capacity may be restored, and the offender may be less likely to recidivate.

For example, we think drug-addicted offenders are often best handled by special drug courts. Roughly half of the over 2,800 drug courts in the United States offer medication under the supervision of a health-care provider as rehabilitative treatment (Matusow et al. 2013). On our executive view, addiction treatment may reduce the strength of persistent, intrusive cravings for an illegal drug that may overwhelm executive processes and result in diminished capacity responsibility. Medication is most effective when given in conjunction with therapy, which assists the drug-addicted offender in learning new, healthier dispositions to act and helps them learn to avoid triggers for their addiction (Bahr, Masters, and Taylor 2012). Often offenders are motivated to participate in rehabilitative programming because they are asked to plead guilty to the charges for which they were arrested but the normal penalties for the charges are held in abeyance during treatment (Bahr, Masters, and Taylor 2012). If the offender finishes his or her treatment successfully, the charges are dropped; if the offender fails to finish treatment, he or she is sent to jail or prison to serve his or her sentence (Bahr, Masters, and Taylor 2012).

Where necessary, an enriched prison environment can serve both retributive aims (by denying offenders important liberties) and rehabilitative aims (by reducing the likelihood of recidivism). Such enriched environments may offer programs within a jail or prison setting that afford inmates opportunities to exercise responsible choice making and to strengthen their practical reason/executive functions. As an example, we would like to offer several programs run out of the second largest jail in the United States, the Cook County Jail in Chicago. Sheriff Tom Dart, head of the jail, has embraced multiple opportunities for inmates to strengthen their cognitive capacities and to develop their agency. Sheriff Dart has initiated a chess program in the jail where inmates can learn to think hard before they make a move, how to plan ahead, and how to recover from their mistakes (Koeske 2016); he has given inmates opportunities to learn useful job skills in construction and given them an opportunity to help their communities by tearing down abandoned buildings (Associated Press 2013); he has helped inmates rethink their relationship with food and the natural environment by allowing them to garden and sell the products at a local farmers' market (Matthews 2015); he allows inmates to learn how to cook gourmet pizzas from a local chef which other inmates may have delivered to their cells as a part of the "recipe for success" program (Janssen 2017); he gives inmates the opportunity to learn the art of beekeeping to produce their own honey (Bash 2014); he has introduced yoga and dance programs; and he allows inmates to earn their high school diploma and GED while in jail. Some of these programs allow jail inmates to exercise and gain job skills; others allow them to make meaningful choices such that they have opportunities to use and strengthen their practical reason and thus, we think, their executive capacities. Many programs offer inmates a direct opportunity to develop their moral agency. And all of these programs give inmates the chance to protect and even expand their moral agency by allowing them become better choice makers.

Many of these programs can be run relatively cheaply. Yoga, chess, and meditation require little more than an instructor, a few cheap supplies, and instructional space. Sheriff Dart often relies upon volunteer or externally grant-funded instructors. Other programs generate income such that they can pay for themselves, such as gardening, making pizzas, and tearing down abandoned properties. Because the programs are offered within a jail setting, they are probably not as effective as they would be if offered in a prison: Jail inmates are released if they make bail, their case is dismissed, they plead out, or their case is decided via a trial. If these programs were available to prisoners with longer sentences and set dates of release, they might have a larger impact, and if they were offered with incentives for participation, such as good time credits, they might be popular with inmates.

Finally, we think solitary confinement should not be utilized unless absolutely required (only where an inmate is an immediate, direct, and serious threat to himself or others). In the rare cases where solitary is used, it should be imposed on inmates for very short time periods. The aim of retribution does not justify the psychological harm of complete isolation and denial of choice making and agency unless there is immediate threat to safety.

Conclusion

Criminal offenders are blameworthy based upon the capacities they possessed at the time harm was caused, and these capacities, in addition to possessing certain mental states causally linked to harm, can serve as the basis for desert (the notion that the person deserves some sort of response to the harm). We claim that for an act to be worthy of praise or blame an actor must possess an MWS at the time of the act. The amount of praise or blame depends on the appropriateness and level of involvement of the MWS with regard to the action. In general, the greater the involvement of the executive with regard to the harm caused, the more blameworthy an actor is for the harm.

Our understanding of the hierarchy of blameworthiness coheres with the structure of mens rea requirements in the MPC, where the most serious crimes such as first-degree homicide require the mental states of "purposely" or "knowingly" causing the criminal harm (American Law Institute 1985). Excuses to culpability, including mental illness, intellectual disability, or juvenile status, may serve to downgrade the blameworthiness of an offender, ideally resulting in downgrading of the charged offense, where an offender lacks an MWS at the time of his or her crime.

We claim that the U.S. model of punishment is best understood as an example of "limiting retributivism" (Morris 1974), where the upper limit of justified punishment is based upon the level of blame an offender deserves for a particular harmful act (taking into consideration the severity of the harm caused). Within this limit, other forward-looking aims may be pursued. However, many U.S. jurisdictions do not adhere to an important principle of limiting retributivism, the principle of parsimony, and thus assign criminal punishments that are too harsh. Very long sentences of incarceration may be unjust not only because they are disproportionate, but also because they unnecessarily undermine moral and legal agency. Further, they may have a criminogenic effect. Current sentences of incarceration offer little opportunity for nonroutine reasoning and decision-making, and thus prisoners' executive functioning can languish because of underuse. Wherever possible, special courts and community-based sanctions ought to be utilized, and incarcerated prisoners ought to be given ample opportunity to develop their executive functioning via programs that allow them to make decisions, learn and practice new skills, and perform physical exercise.

11 Epilogue

The inquiry that produced this book sprang from a relatively simple question: *What is it about us that enables us to be responsible, and to be held responsible?* It must be something in our minds since certain persons with severe mental illness or cognitive disabilities seem to lack it, since children have not yet developed it, and since older people can gradually lose it. As modern neuroscience has begun to reveal the way our mental capacities are generated by brain structures and their functions, seeking out the responsibility-relevant facts about the mind involves seeking out facts about the brain. That *something* in our minds that enables us to be responsible is the way we control our mental activities and allow them to lead to behavior. Or so we have argued here. The capacity to be reasons-responsive—to understand moral and legal rules and to be able to follow them—is grounded primarily in the brain's executive functions.

Returning to the Cases

We began chapter 1 of this book by describing three cases where assigning responsibility seems especially difficult: the case of former Ugandan child soldier Dominic Ongwen, who went from being abducted as a small boy to allegedly overseeing atrocities committed by the Lord's Resistance Army; the case of Norwegian mass killer Anders Breivik, who murdered 77 innocent people in a bizarre and delusional act of terrorism; and the case of serial rapist, murderer, and diagnosed psychopath Brian Dugan, who preyed on women and girls in the Chicago area in the early 1980s. Each case has its distinctively problematic aspects, but all involve an offender whose mental state at the time of his crimes calls into question the extent to which he understood, or could control his behavior in light of, the wrongfulness of his actions. Now, having laid out our theory of responsibility in detail, we can return to these cases.

Dominic Ongwen

Recall that Ongwen was charged in 2005 by the International Criminal Court with multiple crimes against humanity, including the murder and enslavement of civilians. He was abducted around the age of ten on his way to school in the Gulu District of northern Uganda by soldiers in the Lord's Resistance Army, a cultlike guerilla force led by the self-proclaimed prophet Joseph Kony. There is no doubt that Ongwen was severely abused and traumatized in the process of his forced training to become a child solider. By 18 he was a field commander, and by 25 he had secured a senior rank in the LRA's high command.[1]

Our theory implies, as we claimed in chapter 8, that adolescents should be considered significantly less responsible for their actions than adults. And child soldiers, who are initiated as witnesses and participants in extreme violence while their brains are still undergoing significant maturation, are less responsible for their actions than normal juvenile offenders. We have argued elsewhere that, because their cases are so often marked by proximity to violence, forcible recruitment and coercive treatment, indoctrination and manipulation, child soldiers are at risk of having their capacities for responsible agency disrupted or disordered well beyond the normal scope of adolescent immaturity (Fagan, Hirstein, and Sifferd 2016; see also Wainryb 2011, Rosen 2009, Happold 2008, Webster 2007, Thomason 2016).

But at what point, if ever, is a young person subject to these disordering factors no longer excused because of them? The crimes for which Dominic Ongwen now stands trial at the ICC occurred no earlier than 2002, when he was 26 or 27 years old. It is therefore reasonable to presume—purely as a matter of neuropsychological development—that at the time of these offenses, Ongwen had an MWS (minimal working set of executive functions), and that he is prima facie fully responsible for his actions during that time. Yet when we consider the distinctive effects of violent trauma and the brutality of Ongwen's forcible recruitment into the LRA, we may legitimately wonder whether Ongwen, even as an adult, might have a claim to diminished capacity because his trauma had significant impact on his executive functions. Ideally, we think the ICC would pursue a clinical assessment, using the best tools of neuropsychology, of Ongwen's present executive capacities, which might shed some light on his executive profile at the time of his crimes.

The Ongwen case brings up a dilemma that has been explored elsewhere in the responsibility literature—whether a person with a very unusual upbringing or moral development is excused even if he or she is nevertheless fully *rational*. In an important paper, Susan Wolf argues that on a real-self theory of responsibility, a man she calls JoJo ought to be excused even if he has the capacity to utilize his second-order desires (or deep self) to review his first-order desires because his deep self is *insane* (Wolf 1987). In Wolf's hypothetical case, JoJo is raised by a tyrannical and sadistic father who teaches him that harming and even killing other people is a good thing. It is not JoJo's fault he was raised by such a person, and we might expect that under very extreme conditions—similar to those Ongwen faced—a child will adopt a system of moral value others would abhor. In such conditions it seems odd to blame the resulting adult for his or her bizarre moral system, if we can even call it that.

However, as noted above, we think JoJo and Ongwen are only excused from responsibility if their tragic childhoods led to their having diminished mental capacity—in our terminology, if they lack an MWS. Possession of an MWS would entail the capacity to understand moral and legal rules and to behave in accordance with such rules even if they *didn't care to*. That is, perhaps neither JoJo nor Ongwen feels guilty about breaking moral rules because their affective responses have not been trained to reflect societal and legal norms. But while this may be an explanation of their behavior, it is not an excuse. Most people will at some point in their lives decide to obey an important moral or legal rule when they would really rather not. We obey moral and legal rules for many reasons: certainly, because the thought of harming others makes us feel bad, but also because we don't want to lose our jobs, or get arrested, or harm our reputation, and so on. High-functioning people with autism, for example, seem to be able to correct for diminished affective information by following sets of rules. This capacity is grounded squarely within an agent's MWS. Thus, we feel that persons who possess an MWS, and are exposed to the moral and legal rules of their society, can and should use their executive functions to follow such rules.

Possession of an MWS over time may also allow a person like JoJo to reject the values of his father and slowly habituate dispositions that would make him more likely to be obey moral and legal rules. For example, JoJo might slow down his decision-making processes and try to take the perspective of others or learn to better manage his anger and aggression by way of anger management therapy. We understand when a young child raised by a racist mother is racist; but after the child becomes an adult, we blame the former child when they haven't used their rational capacities to try to change attitudes that they should know are wrong.

Adult executive processes have the power to control and alter the habituation processes, as we noted in the foregoing chapters, especially in chapters 3 and 4. But at what age they tend to attain this power is not well known. Once Ongwen reached adulthood, he may have possessed the executive capacity to realize he is causing great harm and begin to reason about whether he should continue his violent life, for instance by noticing the contradiction between his treatment of his fellow soldiers and his treatment of his alleged victims. Then he should have begun to inhibit violent acts, something that did not happen. Why did his capacity to reason and inhibit fail in the ways that they did? Why did they fail while his abilities to plan remained intact? Until we know the answers to these sorts of questions, we won't be able to assess the degree of responsibility that Ongwen bears for each of the crimes he has been charged with.

Anders Breivik

A similar assessment, we believe, should be pursued—or should have been more carefully pursued—in the case of Anders Breivik. Breivik was given two divergent psychiatric diagnoses. He may well have suffered from a severe psychotic disorder, but we still have frustratingly little clinical information about the intactness of his executive functions at the time of his crimes. And that intactness, far more than whether he suffered from persistent psychotic delusions or "only" a personality disorder, is crucial to understanding his fitness for blame and punishment. Without more scientific evidence of his mental condition, we must fall back on what we know about Breivik's horrific crimes. Even if we grant that Breivik's motivating ideas were pathologically delusional, his crimes were the product of extensive, meticulous planning, and he executed them with chilling levels of control and efficiency.

In chapter 9 we argued that all jurisdictions should utilize the MPC test for legal insanity over the common-law M'Naghten test. Under M'Naghten a defendant who recognizes his or her act as criminal or wrong, but lacks the capacity to inhibit an act based upon this knowledge, is still fully culpable for that act. The M'Naghten test for legal insanity excuses a defendant who, due to a severe mental disease or defect, is unable to appreciate the nature and quality of the wrongfulness of his or her act. It is a purely cognitive standard, lacking any volitional or control-based factors. The MPC requires a defendant lack "substantial capacity to either appreciate the criminality of his conduct or to conform his conduct to the requirements of the law" (MPC 4.01). An inability to inhibit action is an executive failure significant enough to indicate that a defendant lacks an MWS. From what we can discern, it seems unlikely that Breivik committed the crime as the result of an "irresistible impulse"—that is, it is unlikely that he would have committed murder with a policeman at his elbow.[2] In the absence of a full clinical evaluation of his executive functioning—which would be welcome—we find it difficult to believe that Breivik lacked an MWS during the time he planned and executed his crime. He therefore lacks a plausible claim to the status of legal insanity, as we think that status is best understood. And lacking any other avenue of legal excuse, Breivik would seem fit to be held fully responsible for his deeds.

Certain executive capacities should be of particular interest in Breivik's case. The ability to detect and resolve conflicts between his current delusional beliefs and his premorbid beliefs should be evaluated. Similarly, we need to know why his inhibitory capacities failed to stop the delusions from forming into concrete plans that were then executed with horrible efficiency. The fact that his plans were well formed and executed obviously loomed large at his competency hearing, but this may just indicate that not all of his executive functions were compromised, something we noted in chapter 2 can happen. We also noted that sometimes other executive processes can correct for a missing one. We need to know whether Breivik could have done this in order to assess his responsibility.

Brian Dugan

Similarly, we believe clarity and justice in Dugan's case could be better secured by the development, testing, and application of finer-grained diagnostic tools for probing executive functions. A diagnosis of psychopathy, like a diagnosis of schizophrenia or bipolar disorder, may indicate to a criminal court that further testing of the defendant's executive functions is warranted. However, we acknowledge here, as we did in chapter 2, that current tools for testing executive functions are coarsely grained. Better tests should be developed to establish more rigorously whether a person has an MWS.

In chapter 9 we indicated that the diagnostic used for psychopathy—the PCL-R—captures a heterogeneous group with regard to executive functions. Psychopaths can differ widely along both executive and affective axes, and the interplay of these factors should be understood as highly relevant to capacity responsibility. If some psychopaths truly qualify as legally insane or subject to diminished capacity, they must lack the executive capacities to control their actions, or to correct for any affective deficits. Classifying Dugan as a psychopath and pointing out his neurological abnormalities does not necessarily indicate such executive incapacity, and so does not mitigate his culpability. Even if Dugan evinces the affective impairments constituting "moral blindness" (Talbert 2008), and even if it is fair to presume that Dugan was similarly impaired during the time of his crimes, those facts *by themselves* would not tend to reduce his culpability. A psychopath with an intact suite of executive functions should still be capable of conforming his or her behavior to moral and legal norms, even if his or her emotional processing of such norms was pathologically disordered.

Recall that Dugan reported being "driven by some kind of an impulse that kept growing" (Gutowski and Mills 2014) and being unable to stop himself. But why exactly? Why didn't inhibitory processes stop his abominable plans from unfolding, or stop him from devising them in the first place? He committed murder and rape frequently enough to have made those behaviors a habit. The executive processes should have the power, we note several times in the previous chapters, to control habituation processes. Why did this not happen? As in the Breivik case, Dugan is more culpable if his other executive processes, such as what appears to be intact planning, could have corrected these deficits. In the end, it seems likely that Dugan's capacities for planning and inhibition were intact, and this is how he eluded arrest for so long. Thus we think it is likely he possessed an MWS, and is responsible for his actions.

Once again, here is our basic schema of criminal responsibility:

Jo is responsible for a criminal illegal act, omission, or consequence only if

1. Jo has a minimal working set of executive functions (MWS), and
2. Jo performed the act or omission, and/or caused the consequence, or failed to act to prevent it, and

3. Jo's executive processes either played an appropriate role in bringing about the action, omission, or consequence, or should have played an appropriate role in preventing it.

Ongwen, Breivik, and Dugan all plausibly satisfy our second criterion for responsibility. They performed a criminal act or acts. We have argued that the responsibility of all three men may turn on whether they meet our first criterion—whether they have an MWS. One of the reasons why these cases seem particularly difficult is because the level of executive functioning of the men is not clear. Whether Ongwen, Breivik, and Dugan possess an MWS—or did at the times of their alleged crimes—also has implications for our third criterion. Assuming the men do possess an MWS, do the crimes of the three involve (or fail to involve) the appropriate sorts of executive activity to satisfy criterion 3? We, of course, cannot know exactly what happened in their minds, but it would appear that each of the three planned at least some of their actions. And we know that these plans were largely effective, which may be evidence that they were intelligently devised with lethal intent. This signals that executive processes did play a role in the actions, and criterion 3 may be met.

There is also a massive failure on the part of all three men to employ empathy and mind-reading capacities to their potential and actual victims. In the latter chapters of the book, we argued for a concept of *executive approval*, which takes place when plans, intentions, and so on have the right sorts of causal contact with the executive processes. This contact occurs as part of the normal operation of the brain's global workspace system. Because of its massive interconnectivity, the workspace system can allow memories, emotions/values, and perception to all have a say in whether an intention or plan of action should go through. This makes the workspace system a good candidate for being the mechanism that generates second-order desires from first-order desires (as Levy 2014 notes, viii–x), or agential approval of an action.

Another way to gain inroads into these three hard cases, then, may be to ask whether it is reasonable to say that someone should employ mind-reading capacities as part of the executive approval process. An agent's ability to do this may depend on their cognitive development. Several factors go into determining the age of adulthood and full responsibility. One of the assumptions behind the entire enterprise of responsibility assessments is that it is possible to specify an age at which both a person's cognitive/

volitional faculties and his or her moral sense have reached a level of development where that person counts as a self-governing moral agent. As we argued chapter 8, a person who has reached the age of say, 22, and has not learned that the lives, plans, and feelings of other humans have a value that should be considered in decision-making bears some responsibility for that situation, and this responsibility is capable of transferring to specific crimes. Because of this, assuming each man had an MWS, we think it is broadly appropriate to assign responsibility to the men for crimes committed after they reached the age of 22.

Concluding Thoughts and Further Questions

Many current theories tend to focus entirely upon synchronic conditions (that is, conditions in effect at the time the crime was committed) as necessary to responsibility. Two popular theories of responsibility claim that an agent is responsible for harm caused by an act that either (1) was under the agent's self-control (in that he or she was "reasons-responsive") in the moments prior to the act or (2) expressed his or her "deep" agency (was subject to review by second-order desires). According to Neil Levy (2014), both theories of responsibility require that an agent's synchronic conscious states be causally related to the harm caused since unconscious mental states are neither reasons-responsive nor reflective of an agent's deep self, and thus it is conscious agency that really matters to responsible action. In chapters 5 through 7, we argued that, in his recent book *Consciousness and Moral Responsibility*, Levy came close to identifying the crucial brain processes for responsibility but instead confused them with their intimate functional partner, the process of consciousness.

Here we have argued that possession of an MWS grounds responsibility not only synchronically, but diachronically. Even where a person's executive functions were not involved in generating an action synchronically, as is the case with habituated action, or in cases where harm is caused by an omission or a lapse, a person may be responsible because the person possessed an MWS for an extended period before the action. Executive functions allow people a certain level of control over their actions by allowing them to manipulate their environment, habituate traits, and even train themselves to emphasize or diminish particular emotional reactions. This sort of diachronic agency grounds basic desert in some lapse cases where individuals should have utilized their MWS to ensure they met an important standard of care.

Our view has implications not only for who is responsible, and for which sorts of acts, but also for how we justify and apply punishment. Studies indicate that consideration of possible arrest and punishment has little to no deterrent effect (Mendes 2004, Nagin 1998). However, if agents truly have expansive diachronic control over their choices, then to have a deterrent effect, the criminal law ought not simply attempt to influence synchronic decisions to commit a crime (or not), but shape citizens' agency over time such that they are law-abiding. Criminal justice policy ought to consider those factors that diminish a person's ability to exercise top-down diachronic agency by limiting their capacities for diachronic self-control, such as poverty, a lack of education, and poor career options. Persons who are impoverished will find it much more difficult to manipulate their environments and habituate dispositions that result in law-abiding behavior. And criminal punishments should be designed to include rehabilitative measures, aimed at influencing offenders' diachronic agency such that they are less likely to commit a crime once released. Unfortunately, as discussed in chapter 10, current punishment practices in the United States, which frequently employ long prison sentences and offer few rehabilitative options, often have the opposite effect: They tend to hobble offenders' control capacities and ability to form socially desirable dispositions, with the result that they are even *more* likely, post-punishment, to commit a crime.

Our primary goal in this book has been to lay out the executive approach to responsibility in a way that is clear and well supported, but also specific and detailed enough to allow for analysis, testing, and improvement by other interested parties. There is a great deal of work remaining, however, including the following projects:

- Further specifying what constitutes a minimal working set of executive processes.
- Creating tests to assess whether a subject has a minimal working set of executive processes.
- Clarifying how the cognitive control network achieves the different executive processes.
- Devising ways to train executive processes.
- Studying the way in which socioeconomic factors influence the executive functions and diachronic agency.

- Studying the effects of different punishments on executive function.
- Devising therapies or other treatments to correct executive deficits and encourage diachronic agency.
- Clarifying the relations between the executive functions and other concepts that are vital to our conceptions of ourselves as agents, such as autonomy and rationality.

The executive approach to responsibility is a theory of the mental ingredients, the building blocks, of responsibility. Our ability to be responsible is built out of our abilities to plan, task switch, resolve conflicts, monitor, inhibit, and so on. These building blocks produce responsible actions by combining and recombining in various permutations until goals, expectations, and intentions have been satisfied. Knowing what these ingredients are can help us understand how we succeed and fail at being responsible in the specific ways that we do. It can also help us learn how to diagnose executive deficits relevant to responsibility and begin to find ways to help people overcome them.

Our theory opens up a way to push past the initial perplexity presented by the difficult cases discussed above. The theory, which reflects both contemporary neuropsychology and the crystallized folk wisdom reflected in the structure of the criminal law, can help categorize these hard cases. We have refined the basic reasons-responsiveness model by specifying executive functions as the core of our reasons-responsive agency and, thus, can rest responsibility upon capacities that can be identified within a particular person or group. We cannot say decisively whether Ongwen, Breivik, or Dugan should be held fully responsible for their deeds; what we *can* say is what sorts of neuroscientific and psychological evidence would be most relevant to settling that question. And that is progress.

Notes

Chapter 1

1. The account of Ongwen's life and case presented here is owed to Baines (2009) and Baines and Ojok (2008).

2. Documents and details pertaining to Ongwen's case at the ICC can be found at https://www.icc-cpi.int/uganda/ongwen. For recent discussions of the moral and legal challenges raised by the prospect of holding child soldiers responsible, see, e.g., Wainryb (2011), Rosen (2009), Happold (2008), Webster (2007), Thomason (2016), and Fagan, Hirstein, and Sifferd (2016).

3. The details of Breivik's case are drawn from Pidd (2012), Flynn (2012), and Knausgaard (2015).

4. The account of Dugan's crimes is drawn from Frisbie and Garrett (2005); the account of his case's recent developments comes from Hughes (2010).

5. One of the authors of this book has argued that a non-eliminative reductive account of the mind can most easily support realism about folk-psychological states and the existence of causally effective mental states (Sifferd 2014). However, the theory promoted here does not require reductivism about the mind.

6. See, for instance, Pardo and Patterson (2013) and Morse (2006a, 2008a).

7. Recently, Morse has been more optimistic—if that is the word—about the theoretical possibility of bridging the gap, saying, "The brain-mind-action relation is a mystery, not because it is inherently not subject to scientific explanation, but rather because the problem is so difficult" (Morse 2017, 163).

8. The most famous example of such putatively threatening research is probably the work of Benjamin Libet (1992, 1996), whose experiments some saw as flatly disproving the existence of human freedom.

9. For examples of this position, see Cashmore (2010), Pereboom (2013), Greene and Cohen (2004), and Caruso (2016).

10. A partial list would include Ayer (1969), Dennett (1984), Wallace (1996), Moore (1997), Fischer and Ravizza (1998), Morse (2013a), and Vargas (2013).

11. In advancing that claim, this book is the fullest and most systematic expression of a theory we have been developing over the past several years. See, for instance, Hirstein and Sifferd (2011); Sifferd and Hirstein (2013); Hirstein and Sifferd (2014); Fagan, Hirstein, and Sifferd (2016); Sifferd, Hirstein, and Fagan (2017); and Sifferd (2016a).

Chapter 2

1. Figure 2.1 indicates that the prefrontal component of the cognitive control network coincides roughly with the middle temporal gyrus. Other imaging studies tend to show smaller areas of activity (e.g., Cole and Schneider 2007, Vincent et al. 2008)—but those areas are almost always within the larger areas shown in figure 2.1.

2. The primary nodes and organs of this network are the amygdala, anterior insular cortex, pre-anterior cingulate cortex, subgenual anterior cingulate cortex, and orbitofrontal cortex.

3. Existing test batteries for executive functions include the Behavioral Assessment of the Dysexecutive Syndrome (BADS; Baron 2007, Wilson et al. 1996) and the Functional Assessment of Verbal Reasoning and Executive Strategies (FAVRES; Higginson et al. 2017).

4. Bennett and Hacker (2003), for example, pose this objection. We will see in chapter 6 that they are half right. When we refer to someone as part of making a responsibility ascription, we refer both to that person and, at another level, to that person's executive processes. At the very least, those posing this objection cannot accuse us of being "reductionists" since our account includes theirs (i.e., it recognizes the importance of behavior) and more.

Chapter 3

1. Todd Long (2004) provides a helpful summary of Fischer and Ravizza's (1998) view:

An agent's responsibility-relevant mechanism K is moderately reasons-responsive iff

(1) K is regularly receptive to reasons, some of which are moral; this requires
    (a) that holding fixed the operation of a K-type mechanism, the agent would recognize reasons in such a way as to give rise to an understandable pattern from the viewpoint of a third party who understands the agent's values and beliefs; and
    (b) that some of the reasons mentioned in (1.a) are moral reasons; and
(2) K is at least weakly reactive to reasons; this requires that the agent would react to at least one sufficient reason to do otherwise (in some possible scenario), although it does not follow that the agent could have responded differently to the actual reasons; and

(3) K is the agent's own; being the agent's own means "taking responsibility" for K; this requires that the agent
   (a) sees herself as the source of her behavior (which follows from the operation of K); and
   (b) believes that she is an apt candidate for the reactive attitudes as a result of how she exercises her agency in certain contexts; and
   (c) views herself as an agent with respect to (3.a)–(3.b) based on her evidence for these beliefs (153).

2. This move is only convincing if it is also claimed an agent has diachronic control over the sorts of reasons he or she entertains. We discuss this issue later in this chapter.

3. According to Julia Annas, a good moral education will teach a person to feel good about the right actions and bad about the wrong ones. Our emotional and intellectual reasoning capacities are developed by the process of education begun by our parents, teachers, role models, and eventually peers, and then continued in our adulthood by our own ability to critically assess our moral opinions. This process of moral education takes time, and it is a process that is never finished; developing a virtuous character trait—or disposition to act—requires cultivating expertise, as with any craft. This is a point Aristotle makes very explicitly in the *Nicomachean Ethics* (Aristotle 1985, 1103a30), and one that Julia Annas (2011) examines in her book *Intelligent Virtue*. An expert electrician will know how to solve novel problems and will be able to articulate to a novice why a particular solution is the right one (Annas 2011, 19). Similarly, expertise in a character trait, such as courage, will mean that a person can apply the trait in new or difficult situations and later provide reasons as to why they acted in such-and-such way. Rote memorization and rule following is often the beginning stage of acquiring a skill, but expertise exhibits intelligent flexibility. Thus, practical reason is critical to the process of becoming "just by doing just actions, temperate by temperate actions, and courageous by courageous actions" (Aristotle 1985, 1103b).

4. As we saw in chapter 2, there are anatomical reasons why cases of patients neatly missing a single executive process should be rare, if all the executive processes are performed by the same brain network.

5. For example, some neurological patients who are confabulatory show an executive deficit in the realm of memory while not confabulating about perceptions.

6. In contrast, patients with Alzheimer's disease, to give one example, do not make good subjects of study because their brain damage is so widespread, making attempts at localization much more difficult.

7. However, do we really use conscious cognition to plan and cause our own actions? In the 1990s, Benjamin Libet (2009, 1992) performed a series of experiments which he argued showed that our sense that we control our actions is a sort of illusion. In the experimental trials, subjects were told to flex their wrists when they felt like doing so, anytime within the next 20 seconds. There was a large clock-like display visible to the subjects with a hand that moved quickly enough to allow for

accurate timing measurements. The subjects were told to observe where the sweeping hand was when they were "first aware of the wish or urge to act." EEG was monitored during the whole process and revealed something surprising: The brains of the subjects showed activity in a part that controlled wrist movements *before* the subject reported awareness of the urge to act. This made it appear that this prior activity (or at least some sort of prior activity) caused both the wrist movement and the subject's experience of making a decision to move. The decision to move, then, was not causing anything; it was being caused by prior, presumably unconscious, brain events. Does the Libet experiment refute the folk psychology of intentions and how they work to produce actions?

We don't think it does. The experimental situation does include a causally potent act of intention formation, produced by executive processes, we submit. This act occurs when the subject is first instructed about the nature of the experimental trials, prior to the first one. At this point, the subject plans (or intends) to move his wrist at some point (see Dennett 2003). The exact point is not specified by the intention, and indeed one simply finds one's wrist moving at a certain point, rather mysteriously. This somewhat spontaneous movement need not be directly preceded by executive activity. That already occurred when the subject planned to move his wrist *at some point* at the beginning of the experiment. But this movement would not have happened at all, and it would not have happened inside the specified time interval, had the subject not formed that prior intention. Libet's piecemeal conception of the nature of human intention and action is mistaken. Planning and intention formation operate at higher levels and over larger time spans than he imagined. When we go to the ATM to get cash, we spot an ATM and form the intention to use it. There need not be a special act of intention prior to every bodily motion on our part. We don't need to, in the process of entering our password, intend to push the 2 button, then intend to push the 5 button, and so on, then intend to take the money, intend to take the receipt, and so on. All those subactions happen as either habituated actions not requiring separate intentions or as actions that received "executive approval" to occur during the formation of the initial plan and intention to visit that ATM.

8. We trace the presence of the "should" in criterion 3 simply to the fact that responsibility is partly an ethical notion. Our use of "should" rests on what we hope is an uncontroversial assumption: One should try to prevent bad things from happening and obey moral rules, and regarding legal responsibility, one should attempt to be law-abiding.

9. The moral responsibility persons may bear for thoughts or plans themselves—even ones that never lead to intentions to act—is an open and interesting question we cannot pursue fully here. Is the bank teller who fantasizes all day about how exactly he would rob his bank blameworthy for those fantasies? Among other things, it seems to matter how much reality or value one attaches to the thoughts or plans under consideration. When one thinks of committing an act of assault, are

there strong emotions, perhaps reactive emotions, directing the agent to harm another person? Are they mere daydreams, or are they closer to concrete plans? If we could look at the inhibitory systems in such a person's brain, we might be able to tell whether they were permanently disposed to inhibit any such actions or only temporarily disposed to inhibit them.

Chapter 4

1. MPC §2.02 stipulates the General Requirements of Culpability, noting that "a person is not guilty of an offense unless he acted purposely, knowingly, recklessly or negligently, as the law may require, with respect to each material element of the offense." The language defining negligence states that a person acts negligently when "he *should be aware* of a substantial and unjustifiable risk that the material element exists or will result from his conduct. The risk must be of such a nature and degree that the actor's failure to perceive it, considering the nature and purpose of his conduct and the circumstances known to him, involves a gross deviation from the standard of care that a reasonable person would observe in the actor's situation."

2. An obvious example in which an offender might truly have lacked normative competence at the time of a crime, but where this incapacity would not provide a basis for excuse, would be a crime committed while in an alcoholic blackout. In this case (speaking broadly) the offender would be deemed responsible for his or her own incapacity, and that incapacity would therefore fail to excuse the offender from culpability for the harm caused (more on this in chapter 7).

3. We believe our theory can help to determine what sort of mental illnesses could most plausibly ground the excuse known as *legal insanity*; chapter 8 will discuss this topic in some detail.

4. We have been developing this thesis for some time; this book represents its fullest and most mature expression (see also Hirstein and Sifferd 2011, Sifferd 2013, Fagan, Hirstein, and Sifferd 2016, Sifferd, Hirstein, and Fagan 2017).

Chapter 5

1. The first condition mentioned here is criterion 2 (the causal criterion) in our theory of responsibility in chapter 3, the second condition is criterion 3 (the executive functions criterion). Our mention that the person in question must be competent covers criterion 1 (the requirement of a minimal working set of executive processes).

2. We think mental events are just a subclass of physical events, whether physical events are ultimately made of matter, or not. Notice here that we favor the traditional, narrower notion of mental states, according to which they are contained

entirely inside the person's head, as opposed to the more fashionable approach of allowing mental states to include external items, such as computers or cell phones (see, e.g., Clark 2008 or Noë 2010). One problem with this view is that it is not folk-consistent—the folk see the mind as inside the head. In addition, even though we might use a digital device in order to plan a crime, the ultimate locus of responsibility for that crime still resides with our executive processes, which are brain processes located inside the head.

3. Levy describes his view as offering us "the best chance of vindicating those aspects of folk psychology most directly tied to our concepts of agency and responsibility" (Levy 2014, vii).

4. Levy adds that he "will also show that only actions performed consciously express our evaluative agency, and that expression of a moral attitude requires consciousness of that attitude" (Levy 2014, 39–40).

5. See also Block (2007b).

6. "Though the notion is not exactly equivalent to the notion of access consciousness," says Levy, "as a matter of empirical fact the two are coextensive" (Levy 2014, 35). Levy says, "I prefer to avoid Block's terminology, because his definition of access consciousness strikes me as slightly question-begging in the context of debates over the role of awareness in behavior" (Levy 2014, 29).

7. See Hirstein (2012, 107–11).

8. "The kind of consciousness at stake," Levy says, "is a state with an informational content; whether it has a phenomenal content as well (or in virtue of its informational content) does not matter for our purposes.... [I]t is not the mark of consciousness of an informational state that there is such a phenomenal content (from now on, I will use the word 'awareness' and its cognates to refer to this kind of informational state...)" (Levy 2014, 29).

9. "It is direct moral responsibility that is at issue, where direct moral responsibility is contrasted with indirect moral responsibility: an agent is indirectly morally responsible for an action if she is directly morally responsible for an earlier action and there is the right kind of causal relationship between the earlier actions and the later" (Levy 2014, 3; see also 92).

10. Thus, the core grammatical construction, one might say, for Levy is the transitive sense of "consciousness," or the "consciousness of" sense, as in "Sue is conscious of a coffee aroma," rather than the intransitive sense, "Sue is conscious," or the state sense, "Sue has a conscious state," or "There is a conscious state in Sue's brain."

11. Levy cautions that "reportability is a very reliable heuristic for awareness; however, it is not necessary and it may not be sufficient" (Levy 2014, 30).

12. Indeed, it may be that all of that filling in and smoothing out of conscious states is actually for the benefit of the executive processes. Perhaps they need their input carefully prepared (including the addition of visual colors) and predigested for them into a form they can work with (see Hirstein 2012, chapter 4).

13. See Zeki and Bartels (1999) and Lamme (2006). The work of Singer (1993) and Llinás (Llinas and Ribary 1993, Llinas, Leznik, and Urbano 2002) established the initial connection between these oscillations and conscious states. There may not be a neat equation between conscious states and these oscillatory states, however, since there appear to be conscious states without the right sort of oscillations (e.g., Hermes et al. 2014).

14. Or the global theorists are at least lumping the prefrontal components of the cognitive control network in with the conscious state, given the possibility that the parietal activity (at least inferior parietal activity) detected could be the workspace buffer, in some cases.

15. One form the debate between the two theory classes has taken recently is whether there can be consciousness without attention (localists tend to believe that there can be such states, globalists tend to believe that there cannot be such states), see, e.g., Koch and Tsuchiya (2007) and Jennings (2015). One argument frequently given for the claim that attention is required for consciousness is based on a phenomenon called the need for attention to see change (see, e.g., Rensink, O'Regan, and Clark 2003). As in the case of stimuli in the periphery of our visual field, though, the fact that we are bad at noticing differences in stimuli or reporting their properties does not imply that consciousness is not present.

16. Block (2007a) argues convincingly for this point.

17. The resolution of binocular rivalry is another example of the sort of "coherencing" process we mentioned above that operates before visual processing reaches the level of consciousness.

18. Dream states apparently do involve some active prefrontal connections to the anterior cingulate (Maquet 2000, Maquet et al. 2005) but possess no significant connections to the cognitive control network since two of its primary processing nodes, the dorsolateral prefrontal cortex and the (superior) parietal cortex, are shut down during REM sleep.

19. We would alter the phrasing of this in a couple of ways, neither of which impacts the basic claim that prefrontal activity is not necessary for consciousness. As we noted previously the term "neural correlates of consciousness" is misleading in that the conscious states aren't just correlated with the hot-zone activity but *are* the hot-zone activity. We also do not think it is correct to call conscious states with no accompanying executive activity "experiences" for reasons we will make clear in this chapter and the following two (roughly, there is no subject of experience).

20. The error may have its roots in the long history of behaviorist influence in both psychology and philosophy, which also chronically confused mental states with the behavior by which we gain knowledge of them. Levy notes that the evidence that he cites for the global workspace comes from experimental paradigms in which "'consciousness' ... is indexed by report" (Levy 2014, 47). Hulme, Friston, and Zeki (2009) also attempt to separate report mechanisms from the conscious state, in defense of Zeki's microconsciousness view.

21. That is, in some of the experiments, the (inferior) parietal activity reported may be the workspace buffer, being worked on by the allied prefrontal areas.

22. If one is going to include the executive functions in what falls under the concept of consciousness, what is the argument for excluding the other recipients of the broadcast, i.e., the entire workspace system, the memory stores, working memory, the brain's emotional and value systems?

23. One argument against this is that, as we noted above, they are actually talking about something more minimal than that, if phenomenal consciousness is a type of experience, since experience requires a subject, which is not present in the states they describe.

24. It is not completely clear that such subcortical support is needed. See, e.g., Magnin et al. (2010).

25. However, is it not open to Levy to claim that he is using a different concept of consciousness, the one that is commonly picked out by the transitive, or "consciousness of" sense of the word, which occurs frequently in our folk talk? Levy approvingly quotes Block, who says that access consciousness "plays a deep role in our ordinary 'consciousness' talk and thought" (Levy 2014, 29n). If we are willing to equate access consciousness to most or all of our everyday uses of the "consciousness of" sense, Block is right. One problem with this for Levy, though, is that it disconnects his account from large amounts of empirical data, including some that he has relied upon. The majority of neuroscientists working on consciousness are investigating that mysterious thing that binds together input from multiple sources to generate our experience, phenomenal consciousness.

26. Hirstein (2012) suggests that states of access consciousness simply be called states of self-consciousness since they include a conscious state interacting with a self-like entity: the set of executive processes.

27. Can the executive processes operate in the human brain by themselves, without consciousness? There are some indications that they can, but the jury is still out. See van Gaal, De Lange, and Cohen (2012).

28. "I want to claim that consciousness by itself, and not in virtue of the ways in which consciousness makes information available for these further processes, makes a difference in our behavior" (Levy 2014, 63).

Chapter 6

1. See Frege (1956) and Quine (1956) for canonical statements of this problem.

2. It may be that claims of the form "Jack is responsible for giving the boss the finger" inherit their opacity from the action verb, in this case "giving the finger," rather than from the responsibility claim. We noted the chronic ambiguity in such verbs at the start of the chapter. The presence of "responsible" in such a sentence, however, seems to strongly force an intentional reading for "giving," which then increases our sense of the incorrectness of the claim.

3. See McKay and Nelson (2010) for a review of the topic. There is an even older debate about whether sentences with nonreferring noun phrases are false, or meaningless, or something else, but certainly a sentence that purports to report a fact about x when there is no x is seriously flawed and cannot be complete and correct.

4. Another benefit of conceptualizing things in this way is that it allows one to make significant progress on the problem of consciousness, known classically as the mind–body problem. If you are the only person who can ever be aware of your conscious states, this makes them different from every physical state we know about, which all admit of awareness or perception by multiple people. In order to defend physicalism, one must find a way to allow for one person to have direct awareness of another's conscious state (or attempt to argue that conscious states are still physical despite this pronounced epistemic asymmetry with known physical states, e.g., Searle [1992]). Hirstein (2008, 2012) argues that this can be achieved by connecting one person's executive functions to another person's conscious state. In terms of the brain, one would connect one person's cognitive control network to the "hot zone" identified by Koch et al. (2016) located in the other person's brain. Now, when we say, "Jill is aware of Jack's conscious state," our utterance is complete and correct. Tacit reference to Jill's executive functions is achieved, and they are correctly connected to Jack's conscious state. If making the separation between consciousness and the executive processes can move us forward on the mind–body problem, this would seem to constitute an argument for doing so.

5. Crimmins and Perry (1989) use "concrete" or Russellian propositions in their theory. Such propositions, which consist simply of arrangements of physical objects, are useful for physicalists like us since they fit into our sparse ontology. See Cappelen and Lepore (2005) for a view opposing the idea of unarticulated propositional constituents.

6. Crimmins and Perry (1989) do not speculate about what brain processes might be behind the remarkable ability to make and comprehend tacit references to mental representations. On their account, whatever mental faculty is "listening" to the attribution and using its words to assign tacit references to mental representations

is not "hearing" the entire sentence, just certain parts of it. Or, if it is "hearing" the entire sentence, it is not constructing a mental equivalent of all of it, an understanding of the entire sentence, as it were. For instance, it only "hears" the "Mark Twain" part of "Jen believes that Mark Twain is a writer." Contrary to this, on the account we are describing here, this "listener" "hears" and processes the entire sentence, which it (or some closely allied mechanism) is able to use to weigh in on the truth value of the entire claim, "objecting" if it processes any part of the claim as incorrect or incomplete by producing a subtle feeling of wrongness in us.

7. The study of the role of the default mode network in language production and comprehension is new, but there are some encouraging findings. For instance, the left angular gyrus is part of both the default mode network and what researchers call "the semantic network" (Seghier, Fagan, and Price 2010). The study of Greicius and Menon (2004) also appears to show that the default mode network is persistently active unless the task is challenging enough.

8. Baars (1996) appears to be the first one to make this point. See also Baars, Ramsøy, and Laureys (2003); Hirstein (2011, 2012, chapter 8); and Hirstein and Sifferd (2011).

9. There may also be cultural/linguistic differences. In Spanish, for instance, the word for "I"—"yo"—is often omitted, and there are other languages that do not have a first-person indexical term such as "I."

10. See, e.g., the OCD self-help book by Schwartz and Beyette (1997).

11. Notice also that when someone has died, and possesses no more mental activity, we tend to refer to them using such phrases as "the body of John Rogers" rather than "John Rogers."

12. It may be that, just as tacit reference by the content sentence (the "x" in "Jen believes that x") of attitude ascriptions is associated with opacities, tacit reference by the noun phrase of such reports is associated with opacities in the noun phrase of such reports, the "Jen" part. See Richard's (1990) phone booth case. Hirstein also offers some possible cases of such *subject opacity* (Hirstein 2012, 177–82).

13. Levy (2014, 37), our italics. Another statement of the thesis is on p. 39.

14. Levy (2014, 29), our italics. In a footnote at the bottom of p. 31, Levy writes, "Mental states are personal when they can rightly be predicated of the person himself ('Jenni believes that *p*,' for instance); and subpersonal when they can only be predicated of some lower level components of the mind ('edge detection occurs in V1,' for instance)."

15. Levy (2014, 33), our italics. He continues: "Information is personally available, as I shall use the phrase when *the agent* is able to effortlessly and easily retrieve it for use in reasoning *and* it is online." Our italics, except for "and."

Chapter 7

1. It is possible that, because of his continued switching of dates, Bert was never fully conscious that the particular weekend he went to Vegas was his weekend to care for his children. However, it was his weekend, and this information would have been clear to him if he had thought to review the schedule.

2. Levy (2014) aims to provide an account of backward-looking responsibility only. This may be one reason why his theory is less adept at handling lapse and negligence cases: Forward-looking aims of punishment seem particularly important in justifying holding persons responsible for violating reasonable standards of care.

3. The MPC, discussed in chapter 4, provides a comprehensive review of criminal offenses and punishments. It was first published by the American Law Institute in 1962 for use by state legislatures to update and standardize the penal law of the United States of America.

4. MPC §202(2)(d).

5. Levy agrees. In a personal correspondence, he noted that his "project is to argue from the neuroscience to the claim that our intuitions are very frequently wrong, with regard to backwards looking responsibility. I expect and welcome an inability to vindicate folk intuitions" (e-mail communication to the authors, October 22, 2017).

6. Information about the Parks case comes from Broughton et al. (1994) and Brogaard and Marlow (2012). Brogaard and Marlow make a good case for Parks's guilt. Their best argument is that the killing scene shows evidence of a prolonged struggle, and the couple's children testified they were screaming at Parks to stop, making the claim that Parks somehow slept through all this implausible. Parks also had a conceivable reason for wanting his in-laws dead. He had developed a serious gambling problem and had taken money from his family savings, and later embezzled from his job, to cover his losses. He was planning to tell his in-laws about all this the next day. If Brogaard and Marlow are right, then we just need another case in which the actions of the person with REMBD are less prolonged or meet with less resistance.

7. Parks does appear to fit our second criterion, the one requiring that the person cause the death and the harm in the appropriate way. Parks did this, and it would appear to anyone observing his actions during the whole episode that it was done with intent and perhaps even some planning.

8. One might wonder whether we even need to worry about criterion 3 since failing criterion 1 is enough to excuse Parks from responsibility on our view. However, we still need to show how well our criterion 3 performs against Levy's conscious awareness criterion. There also may be cases in which we are uncertain whether the person

involved has a minimal working set, and analysis of whether and how exactly the person fails or passes criterion 3 can shed light on the broader issue of competence, as captured in criterion 1.

9. It is also important to note that Parks's lacking a minimal executive set at the time was not due to any voluntary act on his part. That is, he did not in any reasonable way cause himself to go into that state (see the discussion of our criterion 1 in chapter 3, as well as the discussion below about crimes committed while in a state of alcoholic blackout).

10. Assuming that Parks is not regularly reasons-responsive during his REMBD episode, which seems right to us, Fischer and Ravizza would also be committed to claiming that Parks is not responsible. We would point out here that the reason why Parks is not reasons-responsive is that his executive processes are shut down. And as we noted in chapter 3, we may be able to tell in some cases that a person with nonfunctioning executive processes fails to be reasons-responsive, even without knowing that person's behavioral history.

11. A possible exception is that portions of the anterior cingulate are active during dreaming. The images of dreaming brains show activity confined to the classical boundaries of the cingulate sulcus (e.g., Maquet et al. 2005, Desseilles et al. 2011), whereas the images from cognitive control studies (of awake people) tend to show active areas that include a portion of the anterior cingulate, but also extend dorsally well into the supplementary motor area, as in our figures 2.1 and 2.2. Shackman et al. (2011) offer results showing that the portion of the dorsal anterior cingulate active during cognitive control tasks coincides with an area active during pain and negative affect, opening up the possibility that the anterior cingulate activation seen in dreams is due to negative emotions rather than cognitive control (especially since the rest of the cognitive control network is deactivated).

12. See, e.g., Siclari et al. (2017, 2013) for interesting recent attempts to find the neural locus of dreaming.

13. To bolster his case, Levy offers "evidence that somnambulism is a disorder of consciousness" (Levy 2014, 73): A study of a sleepwalking patient "found deactivation of large areas of frontal and parietal association cortices during the episode, together with activation in anterior cingulate cortex and thalamus. In short, some of the brain regions associated with consciousness were active, others were not" (73). Levy goes on to say that "fronto-parietal impairment has also been found in other global automatisms, such as absence seizures and complex partial seizures" (74).

But again, the frontoparietal deactivation indicates that the cognitive control network is shut down, not that there are no conscious states, as the arguments against the globalist theories in chapter 5 appear to show. Indeed, as we noted, that particular profile—prefrontal lobes and parietal areas deactivated, with continued anterior cingulate activity—is what is found in normal REM sleep. It indicates a nonfunctioning cognitive control network, not an absence of consciousness.

14. Apparently, the conscious states in Parks's brain had some sort of representational or intentional status, in that they were not treated by the brain as mere fantasies. Did they have more than the level of representational status that dreams normally have? The conscious states of the REMBD sufferer appear to be an odd mix of internally generated contents and perceived contents. Parks opened a door, walked into a room, grabbed a knife, and stabbed in an effective way, all of which require actual perception. Consciousness is required for flexible behavior, but there can be consciousness without flexible behavior, as in the case of driving on autopilot. Parks's behavior has little to no flexibility, but we suggest that there are still conscious states in his brain.

It is also important to note the strong similarity between people who can only follow action scripts and the dysexecutive patients we described in chapter 2 who showed utilization behavior, e.g., by using whatever tool they see. This sort of behavior is associated with lack of executive processing, not a lack of consciousness.

15. Interestingly, Levy sometimes uses the phrase "Parks himself" to distinguish the waking Parks from the dreaming version. We might, in the language of our theory, describe this as a technique to emphasize tacit reference to the executive process. Levy says, "There is a wide range of reasons which Parks himself would (were he conscious of them) take to be sufficient reasons to inhibit or modulate his behavior…" (Levy 2014, 113). We use this construction too, for the same purpose, but we are able to justify and explain the terminology.

16. Note that, at the other end of his addiction, the beginning, Pete may well have shown clear executive involvement, and hence increased responsibility, for instance, by knowingly and willingly planning how to obtain the drug.

17. There is interesting recent evidence showing that dreams may involve the default mode network. See, e.g., Fox et al. (2013).

Chapter 8

1. Alfred Mele, in the specific context of moral responsibility, puts this question this way: "How does an agent come to be morally responsible for anything?… More fully, how do we get from being neonates who are not morally responsible for anything to being the free, morally responsible agents we are now, if we are indeed free and morally responsible agents?" (Mele 2006, 125).

2. The most widely known of those experiments, reported in, e.g., Mischel, Ebbesen, and Raskoff Zeiss (1972), gave young children a choice: eat one preferred treat immediately, or wait several minutes and earn the chance to have *two* treats.

3. For research on the emergence of moral emotions in children, see, e.g., Barrett, Zahn-Waxler, and Cole (1993); Zahn-Waxler and Robinson (1995); Barrett (1998); Vaish, Carpenter, and Tomasello (2016).

4. Bensouda's statement, given on December 6, 2016, can be viewed on YouTube; a transcript is available at https://www.icc-cpi.int/Pages/item.aspx?name=2016-12-06-otp-stat-ongwen.

5. For a recent study on the (in)effectiveness of juvenile transfer as a deterrent, see Zane, Welsh, and Mears (2016).

6. The Eighth Amendment states, in its entirety, "Excessive bail shall not be required, nor excessive fines imposed, nor cruel and unusual punishments inflicted."

7. Roper v. Simmons, 543 U.S. 551, 125 S. Ct. 1183, 161 L. Ed. 2d 1 (2005); Graham v. Florida, 130 S. Ct. 2011, 560 U.S. 48, 176 L. Ed. 2d 825 (2010); Miller v. Alabama, 132 S. Ct. 2455, 567 U.S. 460, 183 L. Ed. 2d 407 (2012); Jackson v. Hobbs, 132 S. Ct. 1733, 182 L. Ed. 2d 251 (2012).

8. In discussing the development of executive functions, we are roughly employing the integrative, "unity with diversity" account of executive functions favored by, e.g., Miyake and Friedman (2012) and Diamond (2013). This framework was originally developed as a model of adult executive function—leading some to wonder about its applicability to developmental contexts (see Garon, Bryson, and Smith 2008)—and it is not universally endorsed, but it has shown genuine stability as the dominant theoretical view in the field.

9. "Metacognition" is, roughly, the ability to think *about* one's own thinking, which is of obvious importance to higher-level reasoning.

10. See, for instance, Annas (2011), Vincent (2009), Annas (2003), Driver (2001), Arpaly (2003), Sifferd (2016b), and Webber (2006).

11. Even putting aside the not-insignificant number of philosophers who, for various reasons, are skeptical about human responsibility, period, e.g., Caruso (2016), Pereboom (2013), and Levy (2014).

12. In fairness to Fischer and Ravizza, they do acknowledge that children undergo a process of development by which they become morally responsible—one that first involves a kind of pretense, by which we treat children *as if* they were responsible for the purpose of moral training. This leads to children's learning to *take* responsibility for their actions. As this process iterates and the child grows into an adult, the child's reasons-responsive mechanism develops the right kind of history, and the child becomes *ready* to be held responsible (Fischer and Ravizza 1998, 208–10). Describing this gradual process by which one attains responsibility, though, is not a substitute for a genuinely scalar conception of responsibility.

13. Surveyed earlier in this chapter. For representative reviews, see Lahat (2015); Carlson and Moses (2001); Best, Miller, and Jones (2009); Blakemore and Choudhury (2006); Garon, Bryson, and Smith (2008); and Diamond (2013).

14. The terminology of "cool" tasks or contexts refers to those with lower emotional or motivational salience, as contrasted with more emotionally significant "hot" tasks (Prencipe et al. 2011).

15. We will return to the topic of punishment in chapter 10.

Chapter 9

1. It may further one specific purpose of the criminal law to *incapacitate* the defendant, namely, the purpose of incapacitation. If a person is unable to keep from behaving dangerously despite recognizing the illegality of such actions, then there are good forward-looking reasons to mitigate the harms such a person can do (which may be served by detention in a treatment facility or the like). However, taking such measures does not necessarily imply an attribution of responsibility, and it is hard to see how aims of justice or deterrence, for instance, would be served by assigning responsibility to a person who lacked, and may continue to lack, "a fair opportunity to avoid wrongdoing" in Brink's (2013, 8) terms.

2. As of this writing, 25 U.S. states employ some version of the M'Naghten rule, compared to 20 states (and the District of Columbia) using some version of the MPC rule; five states either use a different standard or have abolished the insanity defense altogether (FindLaw, Redding 2006).

3. A fuller discussion of the general notion of normative competence can be found in chapter 4.

4. A sampling of recent work on the executive functioning of patients with schizophrenia would include, e.g., Fucetola et al. (2000), Heinrichs and Zakzanis (1998), Johnson-Selfridge and Zalewski (2001), Krishnadas et al. (2014), Laloyaux et al. (2014), Lee and Park (2005), Hutton et al. (1998), Li (2004), González-Blanch et al. (2006), and Leeson et al. (2008).

5. For empirical studies on the interaction between executive functions and spurious or delusive beliefs, see, e.g., Davies et al. (2001); Langdon and Coltheart (2000); and McKay, Landgon, and Coltheart (2007).

6. The "policeman at his elbow," or irresistible impulse, standard for legal insanity stems from the early common-law case of People v. Hubert, 119 Cal. 216, 223–4 (1897). Because it qualifies as a "control" condition, it bears some similarity to the MPC rule. However, courts that have adopted the MPC substantial-capacity approach have largely rejected the standard as too strict. As the court in People v. Jackson, 245 Mich. App. 17; 627 N.W.2d 11 (2001) noted,

> This is not to say that the "policeman at the elbow" standard has no relevance. If it is approached as being one of many avenues of inquiry, the hypothetical is directly probative of one dimension of a defendant's capacity to control his conduct as required by law. Certainly, if credible testimony offered by a defendant establishes that he could not refrain from acting even if faced with

immediate capture and punishment, then the defendant would have gone a long way toward establishing that he lacked the requisite substantial capacity to conform to requirements of the law. The converse, however, is not true. A defendant who could resist until the threat posed by a policeman had passed does not necessarily possess the capacity to conform. Nonetheless, if it so chooses, the prosecution must be allowed to explore the depths of defendant's alleged incapacity by posing the "policeman at the elbow" hypothetical inasmuch as the question is probative of a defendant's ability to conform to the requirements of the law under the most extreme circumstance of immediate capture and punishment.

7. An encouraging new research direction has found "a common pattern of disruption across major psychiatric disorders" [including psychotic, bipolar and unipolar depressive, anxiety, and substance use disorders] in what they call the "multiple demand network," which is another name for the cognitive control network (McTeague et al. 2017, 676). The authors state that, "within each of the disorder classes impairments are consistent in the multiple demand network across a wide range of cognitive tasks" (McTeague et al. 2016, 37). We think it makes sense that these disorders would all involve problems with executive functions, and that it may be a tacit awareness of this common element that makes us refer to them all as "mental illnesses"—they can disable the highest levels of human function.

8. Listing Gein and Bundy as "real" avatars of psychopathy is not meant to imply that either of these men were psychopaths—only that they are ensconced as such in the minds of many; the popular sense of what psychopathy *is* has been shaped by these high-profile figures who, rightly or wrongly, have been associated with the condition.

9. Those arguing for diminished or absent responsibility include Fine and Kennett (2004); Levy (2007); and Focquaert, Glenn, and Raine (2015); those arguing that (at least some) psychopaths should be considered responsible include Maibom (2008), Sifferd and Hirstein (2013), and Jurjako and Malatesti (2016).

Chapter 10

1. See, e.g., Moore (1997), Duff (1990), Morse (2015), Vargas (2013), Shoemaker (2015), Murray and Nahmias (2014), Nahmias (2014), and Nahmias and Murray (2010).

Chapter 11

1. The details of Ongwen's biography, and his alleged crimes, are drawn from Baines (2009), Baines and Ojok (2008), and the International Criminal Court (2016).

2. As discussed in chapter 9, one way that courts can determine whether a defendant had the ability to conform his or her conduct to the law (one of the prongs of the MPC test) is the irresistible impulse test, which asks whether the defendant would have committed the crime with a "policeman at his elbow."

References

Addie, S., Adams, B., Firestine, K., & Griffin, P. (2011). Trying juveniles as adults: An analysis of state transfer laws and reporting. *Juvenile Offenders and Victims: National Report Series*. U.S. Department of Justice: Office of Juvenile Justice and Delinquency Prevention. https://www.ncjrs.gov/pdffiles1/ojjdp/232434.pdf

Aharoni, E., Sinnott-Armstrong, W., & Kiehl, K. A. (2012). Can psychopathic offenders discern moral wrongs? A new look at the moral/conventional distinction. *Journal of Abnormal Psychology*, 121(2), 484–497.

Aharoni, E., Sinnott-Armstrong, W., & Kiehl, K. A. (2014). What's wrong? Moral understanding in psychopathic offenders. *Journal of Research in Personality*, 53, 175–181.

American Law Institute. (1985). *Model Penal Code and Commentaries: Official Draft and Revised Comments*. Philadelphia: American Law Institute.

American Law Institute. (2007). *Model Penal Code: Sentencing, Tentative Draft No.1, Part I. General Provisions*. Philadelphia: American Law Institute.

American Psychiatric Association. (1994). *Diagnostic and Statistical Manual of Mental Disorders* (4th ed.). Washington, DC: American Psychiatric Association.

Andrews-Hanna, J. R. (2011). The brain's default network and its adaptive role in internal mentation. *Neuroscientist*, 18(3), 251–270. doi:10.1177/1073858411403316.

Andrews-Hanna, J. R., Reidler, J. S., Sepulcre, J., Poulin, R., & Buckner, R. L. (2010). Functional-anatomic fractionation of the brain's default network. *Neuron*, 65(4), 550–562. doi:10.1016/j.neuron.2010.02.005.

Annas, J. (2003). The structure of virtue. In L. Zagzebski & M. DePaul (Eds.), *Intellectual Virtue: Perspectives from Ethics and Epistemology* (pp. 15–33). New York: Oxford University Press.

Annas, J. (2011). *Intelligent Virtue*. Oxford: Oxford University Press.

Aristotle. (1984). *The Complete Works of Aristotle* (Vol. 2) (G. A. Kennedy, Trans.). Princeton, NJ: Princeton University Press.

Aristotle. (1985). *Nicomachean Ethics* (T. Irwin, Trans.). Indianapolis, IN: Hackett.

Arpaly, N. (2003). *Unprincipled Virtue*. Oxford: Oxford University Press.

Associated Press. (2013). Cook County inmates to tear down vacant buildings. *NBC Chicago*. https://www.nbcchicago.com/news/local/Cook-County-Inmates-to-Tear-Down-Vacant-Buildings-236727671.html.

Ayer, A. J. (1969). Freedom and necessity. In *Philosophical Essays* (pp. 271–284). New York: Bedford/St. Martin's Press.

Baars, B. J. (1993). *A Cognitive Theory of Consciousness*. Cambridge: Cambridge University Press.

Baars, B. J. (1995). Evidence that phenomenal consciousness is the same as access consciousness. *Behavioral and Brain Sciences*, 18(2), 249. doi:10.1017/s0140525x00038218.

Baars, B. J. (1996). Understanding subjectivity: Global workspace theory and the resurrection of the observing self. *Journal of Consciousness Studies*, 3(3), 211–216.

Baars, B. J. (2002). The conscious access hypothesis: Origins and recent evidence. *Trends in Cognitive Sciences*, 6(1), 47–52. doi:10.1016/s1364-6613(00)01819-2.

Baars, B. J., Ramsøy, T. Z., & Laureys, S. (2003). Brain, conscious experience and the observing self. *Trends in Neurosciences*, 26(12), 671–675. doi:10.1016/j.tins.2003.09.015.

Baddeley, A. (1998). The central executive: A concept and some misconceptions. *Journal of the International Neuropsychological Society*, 4(5), 523–526. doi:10.1017/s135561779800513x.

Baddeley, A. (2002). Fractionating the central executive. In D. T. Stuss & R. T. Knight (Eds.), *Principles of Frontal Lobe Function* (pp. 246–260). New York: Oxford University Press.

Baddeley, A., & Wilson, B. (1988). Frontal amnesia and the dysexecutive syndrome. *Brain and Cognition*, 7(2), 212–230. doi:10.1016/0278-2626(88)90031-0.

Bahr, S. J., Masters, A. L., & Taylor, B. M. (2012). What works in substance abuse treatment programs for offenders? *Prison Journal*, 92(2), 155–174.

Baines, E., & Ojok, B. (2008). Complicating victims and perpetrators in Uganda: On Dominic Ongwen. Justice and Reconciliation Project Field Note 7, July 2008. http://justiceandreconciliation.com/wp-content/uploads/2008/07/JRP_FN7_Dominic-Ongwen.pdf

Baines, E. K. (2009). Complex political perpetrators: Reflections on Dominic Ongwen. *Journal of Modern African Studies*, 47(2), 163–191.

Banich, M. T. (2009). Executive function: The search for an integrated account. *Current Directions in Psychological Science*, 18(2), 89–94.

Baron, I. S. (2007). Behavioural Assessment of the Dysexecutive Syndrome for Children (BADS-C) by Emslie, H., Wilson, F. C., Burden, V., Nimmo-Smith, I., & Wilson, B. A. (2003). *Child Neuropsychology*, 13(6), 539–542. doi:10.1080/09297040601112781.

Barrett, K. C. (1998). The origins of guilt in early childhood. In J. Bybee (Ed.), *Guilt and Children* (pp. 75–90). San Diego, CA: Academic Press.

Barrett, K. C., Zahn-Waxler, C., & Cole, P. M. (1993). Avoiders vs. amenders: Implications for the investigation of guilt and shame during toddlerhood? *Cognition and Emotion*, 7(6), 481–505.

Barrett, L. F., & Satpute, A. B. (2013). Large-scale brain networks in affective and social neuroscience: Towards an integrative functional architecture of the brain. *Current Opinion in Neurobiology*, 23(3), 361–372. doi:10.1016/j.conb.2012.12.012.

Barron, A. B., & Klein, C. (2016). What insects can tell us about the origins of consciousness. *Proceedings of the National Academy of Sciences of the United States of America*, 113(18), 4900–4908. doi:10.1073/pnas.1520084113.

Bash, H. (2014). Urban farming at Cook County Jail. *Austin Weekly News*, January 14, 2014. http://www.austinweeklynews.com/News/Articles/1-14-2014/Urban-farming-at-Cook-County-Jail/.

Bauer, J., Werner, A., Kohl, W., Kugel, H., Shushakova, A., Pedersen, A., et al. (2016). Hyperactivity and impulsivity in adult attention-deficit/hyperactivity disorder is related to glutamatergic dysfunction in the anterior cingulate cortex. *World Journal of Biological Psychiatry*, 1–9. doi:10.1080/15622975.2016.1262060

Bellebaum, C., & Daum, I. (2007). Cerebellar involvement in executive control. *Cerebellum (London, England)*, 6(3), 184–192. doi:10.1080/14734220601169707.

Bennett, M. R., & Hacker, P. M. S. (2003). *Philosophical Foundations of Neuroscience*. London: Blackwell.

Best, J. R., Miller, P. H., & Jones, L. L. (2009). Executive functions after age 5: Changes and correlates. *Developmental Review*, 29(3), 180–200. doi:10.1016/j.dr.2009.05.002.

Bilderbeck, A., Farias, M., Brazil, I., Jakobowitz, S., & Wikholm, C. (2013). Participation in a 10-week course of yoga improves behavioral control and decreases psychological distress in a prison population. *Journal of Psychiatric Research*, 47, 1438–1445.

Birbaumer, N., Veit, R., Lotze, M., Erb, M., Hermann, C., Grodd, W., et al. (2005). Deficient fear conditioning in psychopathy: A functional magnetic resonance imaging study. *Archives of General Psychiatry*, 62(7), 799–805.

Blair, R. J. R. (1995). A cognitive developmental approach to morality: Investigating the psychopath. *Cognition*, 57(1), 1–29.

Blair, R. J. R. (1997). Moral reasoning and the child with psychopathic tendencies. *Personality and Individual Differences*, 22(5), 731–739.

Blair, R. J. R. (2007). Empathic dysfunction in psychopathic individuals. In T. F. D. Farrow & P. W. R. Woodruff (Eds.), *Empathy in Mental Illness* (pp. 3–16). Cambridge: Cambridge University Press.

Blakemore, S. J., & Choudhury, S. (2006). Development of the adolescent brain: Implications for executive function and social cognition. *Journal of Child Psychology and Psychiatry, and Allied Disciplines*, 47(3–4), 296–312.

Block, N. (2007a). Consciousness, accessibility, and the mesh between psychology and neuroscience. *Behavioral and Brain Sciences*, 30(5–6), 481–499. doi:10.1017/s0140525x07002786.

Block, N. (2007b). *Consciousness, Function, and Representation: Collected Papers*. Cambridge, MA: MIT Press.

Bortolotti, L., Broome, M. R., & Mameli, M. (2014). Delusions and responsibility for action: Insights from the Breivik case. *Neuroethics*, 7(3), 377–382. doi:10.1007/s12152-013-9198-4.

Braver, T. S. (2001). Anterior cingulate cortex and response conflict: Effects of frequency, inhibition and errors. *Cerebral Cortex*, 11(9), 825–836. doi:10.1093/cercor/11.9.825.

Brickner, R. M. (1952). Brain of patient A. after bilateral frontal lobectomy; status of frontal-lobe problem. *Archives of Neurology and Psychiatry*, 68(3), 293–313. doi:10.1001/archneurpsyc.1952.02320210003001.

Brink, D. O. (2004). Immaturity, normative competence, and juvenile transfer: How (not) to punish minors for major crimes. *Texas Law Review*, 82, 1555–1585.

Brink, D. O. (2013). Responsibility, Incompetence, and Psychopathy. The Lindley Lecture, The University of Kansas, Lawrence, KS, April 13, 2013. https://kuscholarworks.ku.edu/handle/1808/12562.

Brink, D. O., & Nelkin, D. K. (2013). Fairness and the architecture of responsibility. In D. Shoemaker (Ed.), *Oxford Studies in Agency and Responsibility* (pp. 284–313). Oxford: Oxford University Press.

Brogaard, B., & Marlow, K. (2012). Sleep driving and sleep killing: The Kenneth Parks case. *The Superhuman Mind.* December 13, 2002. https://www.psychologytoday.com/blog/the-superhuman-mind/201212/sleep-driving-and-sleep-killing.

Broughton, R., Billings, R., Cartwright, R., Doucette, D., Edmeads, J., Edwardh, M., et al. (1994). Homicidal somnambulism: A case report. *Sleep*, 17(3), 253–264.

Brown, D. K. (2002). What virtue ethics can do for criminal justice: A reply to Huigens. *Wake Forest Law Review*, 37, 29–50.

Buckner, R. L., Andrews-Hanna, J. R., & Schacter, D. L. (2008). The brain's default network. *Annals of the New York Academy of Sciences*, 1124(1), 1–38. doi:10.1196/annals.1440.011.

Bunge, S. A., Hazeltine, E., Scanlon, M. D., Rosen, A. C., & Gabrieli, J. (2002). Dissociable contributions of prefrontal and parietal cortices to response selection. *NeuroImage*, 17(3), 1562–1571.

Bunge, S. A., & Wright, S. B. (2007). Neurodevelopmental changes in working memory and cognitive control. *Current Opinion in Neurobiology*, 17(2), 243–250.

Burgess, P. W., Veitch, E., de Lacy Costello, A., & Shallice, T. (2000). The cognitive and neuroanatomical correlates of multitasking. *Neuropsychologia*, 38(6), 848–863. doi:10.1016/s0028-3932(99)00134-7.

Bush, G., Luu, P., & Posner, M. I. (2000). Cognitive and emotional influences in anterior cingulate cortex. *Trends in Cognitive Sciences*, 4(6), 215–222. doi:10.1016/s1364-6613(00)01483-2.

Cappelen, H., & Lepore, E. (2005). *Insensitive Semantics: A Defense of Semantic Minimalism and Speech Act Pluralism.* Oxford: Blackwell.

Carlson, S. M., & Moses, L. J. (2001). Individual differences in inhibitory control and children's theory of mind. *Child Development*, 72(4), 1032–1053. doi:10.1111/1467-8624.00333.

Carruthers, P. (2011). Higher-Order Theories of Consciousness. Last modified August 29, 2016. https://plato.stanford.edu/entries/consciousness-higher/.

Carson, E. A., & Anderson, E. (2016). *Prisoners in 2015.* Washington, DC: US Department of Justice, Bureau of Justice Statistics.

Carter, C. S., Braver, T. S., Barch, D. M., Botvinick, M. M., Noll, D., & Cohen, J. D. (1998). Anterior cingulate cortex, error detection, and the online monitoring of performance. *Science*, 280(5364), 747–749.

Caruso, G. D. (2016). Free will skepticism and criminal behavior. *Southwest Philosophy Review*, 32(1), 25–48. doi:10.5840/swphilreview20163214.

Casey, B. (2015). Beyond simple models of self-control to circuit-based accounts of adolescent behavior. *Annual Review of Psychology*, 66, 295–319.

Cashmore, A. R. (2010). The Lucretian swerve: The biological basis of human behavior and the criminal justice system. *Proceedings of the National Academy of Sciences of the United States of America*, 107(10), 4499–4504.

Christakou, A. (2014). Present simple and continuous: Emergence of self-regulation and contextual sophistication in adolescent decision-making. *Neuropsychologia*, 65, 302–312. doi:10.1016/j.neuropsychologia.2014.09.008.

Cipriani, D. (2009). *Children's Rights and the Minimum Age of Criminal Responsibility: A Global Perspective*. Farnham, UK: Ashgate.

Clark, A. (2008). *Supersizing the Mind: Embodiment, Action, and Cognitive Extension*. Oxford: Oxford University Press.

Coates, D. J., & Swenson, P. (2013). Reasons-responsiveness and degrees of responsibility. *Philosophical Studies*, 165(2), 629–645.

Cohen-Mansfield, J., Shmotkin, D., Blumstein, Z., Shorek, A., Eyal, N., & Hazan, H. (2013). The old, old-old, and the oldest old: Continuation or distinct categories? An examination of the relationship between age and changes in health, function, and wellbeing. *International Journal of Aging & Human Development*, 77(1), 37–57.

Cole, M. W., & Schneider, W. (2007). The cognitive control network: Integrated cortical regions with dissociable functions. *NeuroImage*, 37(1), 343–360. doi:10.1016/j.neuroimage.2007.03.071.

Collette, F., Van der Linden, M., Laureys, S., Delfiore, G., Degueldre, C., Luxen, A., et al. (2005). Exploring the unity and diversity of the neural substrates of executive functioning. *Human Brain Mapping*, 25(4), 409–423. doi:10.1002/hbm.20118.

Colombo, J. (2001). The development of visual attention in infancy. *Annual Review of Psychology*, 52(1), 337–367.

Conway, A. R., & Engle, R. W. (1994). Working memory and retrieval: A resource-dependent inhibition model. *Journal of Experimental Psychology. General*, 123(4), 354–373.

Crick, F., & Koch, C. (1990). Towards a neurobiological theory of consciousness. *Seminars in Neuroscience*, 2, 263–275.

Crick, F., & Koch, C. (2003). A framework for consciousness. *Nature Neuroscience*, 6(2), 119–126. doi:10.1038/nn0203-119.

Crimmins, M. (1992). *Talk about Beliefs*. Cambridge, MA: MIT Press.

Crimmins, M., & Perry, J. (1989). The prince and the phone booth: Reporting puzzling beliefs. *Journal of Philosophy*, 86(12), 685–711. doi:10.2307/2027014.

Cullen, F., Jonson, C., & Nagin, D. (2011). Prisons do not reduce recidivism: The high cost of ignoring science. *Prison Journal*, 91, 48–65.

Curtis, C. E., & D'Esposito, M. (2003). Persistent activity in the prefrontal cortex during working memory. *Trends in Cognitive Sciences*, 7(9), 415–423. doi:10.1016/s1364-6613(03)00197-9.

Damasio, A. (1994). *Descartes' Error: Emotion, Reason and the Human Brain*. New York: Putnam.

Dang-Vu, T. T., Schabus, M., Desseilles, M., Sterpenich, V., Bonjean, M., & Maquet, P. (2010). Functional neuroimaging insights into the physiology of human sleep. *Sleep*, 33(12), 1589–1603. doi:10.1093/sleep/33.12.1589.

Darwall, S. L. (2006). *The Second-Person Standpoint: Morality, Respect, and Accountability*. Cambridge, MA: Harvard University Press.

Davidson, M. C., Amso, D., Anderson, L. C., & Diamond, A. (2006). Development of cognitive control and executive functions from 4 to 13 years: Evidence from manipulations of memory, inhibition, and task switching. *Neuropsychologia*, 44(11), 2037–2078.

Davies, M., Coltheart, M., Langdon, R., & Breen, N. (2001). Monothematic delusions: Towards a two-factor account. *Philosophy, Psychiatry, & Psychology*, 8(2), 133–158.

Dehaene, S. (2014). *Consciousness and the Brain: Deciphering How the Brain Codes Our Thoughts*. New York: Penguin.

Dehaene, S., Kerszberg, M., & Changeux, J.-P. (1998). A neuronal model of a global workspace in effortful cognitive tasks. *Proceedings of the National Academy of Sciences of the United States of America*, 95(24), 14529–14534. doi:10.1073/pnas.95.24.14529.

Dehaene, S., & Naccache, L. (2001). Towards a cognitive neuroscience of consciousness: Basic evidence and a workspace framework. *Cognition*, 79(1–2), 1–37. doi: 10.1016/s0010-0277(00)00123-2.

de la Vega, C., Solter, A., Kwon, S.-R., & Isaac, D. M. (2012). *Cruel and Unusual: US Sentencing Practices in a Global Context*. San Francisco: University of San Francisco School of Law.

Dennett, D. C. (1978). *Brainstorms: Philosophical Essays on Mind and Psychology*. Cambridge, MA: MIT Press.

Dennett, D. C. (1984). *Elbow Room: The Varieties of Free Will Worth Wanting* (Vol. 36). Cambridge, MA: MIT Press.

Dennett, D. C. (2001). Are we explaining consciousness yet? *Cognition*, 79(1–2), 221–237. doi:10.1016/s0010-0277(00)00130-x.

Dennett, D. C. (2003). *Freedom Evolves*. New York: Viking.

D'Esposito, M., Postle, B. R., Ballard, D., & Lease, J. (1999). Maintenance versus manipulation of information held in working memory: An event-related fMRI study. *Brain and Cognition*, 41(1), 66–86. doi:10.1006/brcg.1999.1096.

Desseilles, M., Dang-Vu, T. T., Sterpenich, V., & Schwartz, S. (2011). Cognitive and emotional processes during dreaming: A neuroimaging view. *Consciousness and Cognition*, 20(4), 998–1008.

Diamond, A. (2013). Executive functions. *Annual Review of Psychology*, 64, 135–168.

Doris, J. (2015). *Talking to Our Selves*. Oxford: Oxford University Press.

Driver, J. (2001). *Uneasy Virtue*. Cambridge: Cambridge University Press.

Duff, R. A. (1990). *Intention, Agency and Criminal Liability*. Cambridge, MA: Blackwell.

Duff, R. A. (1993). Choice, character and criminal liability. *Law and Philosophy*, 12(4), 345–383.

Duff, R. A. (2004). Who is responsible, for what, to whom? *Ohio State Journal of Criminal Law*, 2, 441–461.

Duncan, J., & Owen, A. M. (2000a). Common regions of the human frontal lobe recruited by diverse cognitive demands. *Trends in Neurosciences*, 23(10), 475–483. doi:10.1016/s0166-2236(00)01633-7.

Duncan, J., & Owen, A. M. (2000b). Dissociative methods in the study of frontal lobe function. In S. Monsell & J. Driver (Eds.), *Control of Cognitive Processes* (pp. 567–576). Cambridge, MA: MIT Press.

Egner, T., & Hirsch, J. (2005). Cognitive control mechanisms resolve conflict through cortical amplification of task-relevant information. *Nature Neuroscience*, 8(12), 1784–1790. doi:10.1038/nn1594.

Elger, B. S. (2009). Prison life: Television, sports, work, stress and insomnia in a remand prison. *International Journal of Law and Psychiatry*, 32, 74–83.

Fagan, T. K., Hirstein, W., & Sifferd, K. L. (2016). Child soldiers, executive functions, and culpability. *International Criminal Law Review*, 16(2), 258–286. doi:10.1163/15718123-01602002.

Farahany, N. A. (2015). Neuroscience and behavioral genetics in US criminal law: An empirical analysis. *Journal of Law and the Biosciences*, 2(3), 485–509.

Feinberg, J. (1970). *Doing and Deserving: Essays in the Theory of Responsibility*. Princeton, NJ: Princeton University Press.

Field, H. (1973). Theory change and the indeterminacy of reference. *Journal of Philosophy*, 70(14), 462–481. doi:10.2307/2025110.

FindLaw. The Insanity Defense among the States. Accessed December 5, 2017. http://criminal.findlaw.com/criminal-procedure/the-insanity-defense-among-the-states.html.

Fine, C., & Kennett, J. (2004). Mental impairment, moral understanding and criminal responsibility: Psychopathy and the purposes of punishment. *International Journal of Law and Psychiatry*, 27(5), 425–443.

Fischer, J. M. (1994). *The Metaphysics of Free Will*. Oxford: Blackwell.

Fischer, J. M., & Ravizza, M. (1998). *Responsibility and Control: A Theory of Moral Responsibility*. Cambridge: Cambridge University Press.

Flynn, S. (2012). "Is he coming? Is he? Oh God, I think he is." *GQ*. July 30, 2012. https://www.gq.com/story/anders-behring-breivik-norway-massacre-story

Focquaert, F., Glenn, A. L., & Raine, A. (2015). Psychopathy and free will from a philosophical and cognitive neuroscience perspective. In W. Glannon (Ed.), *Free Will and the Brain Neuroscientific, Philosophical, and Legal Perspectives* (pp. 103–124). Cambridge: Cambridge University Press.

Fox, K. C. R., Nijeboer, S., Solomonova, E., Domhoff, G. W., & Christoff, K. (2013). Dreaming as mind wandering: Evidence from functional neuroimaging and first-person content reports. *Frontiers in Human Neuroscience*, 7, 412. doi:10.3389/fnhum.2013.00412.

Frankfurt, H. G. (1969). Alternate possibilities and moral responsibility. *Journal of Philosophy*, 66(23), 829–839.

Frankfurt, H. G. (1982). Freedom of the will and the concept of a person. In G. Watson (Ed.), *Free Will* (pp. 81–95). Oxford: Oxford University Press.

Frase, R. (2004). Limiting retributivism. In M. Tonry (Ed.), *The Future of Imprisonment in the 21st Century* (pp. 83–119). New York: Oxford University Press.

Frase, R. (2009). Norval Morris's contributions to sentencing structures, theory, and practice. *Federal Sentencing Review*, 21, 254–260.

Frässle, S., Sommer, J., Jansen, A., Naber, M., & Einhauser, W. (2014). Binocular rivalry: Frontal activity relates to introspection and action but not to perception. *Journal of Neuroscience*, 34(5), 1738–1747. doi:10.1523/jneurosci.4403-13.2014.

Frege, G. (1956). The thought: A logical inquiry. *Mind*, 65(259), 289–311.

Frisbie, T., & Garrett, R. (2005). *Victims of Justice Revisited: Completely Updated and Revised*. Evanston, IL: Northwestern University Press.

Frith, C. D. (2000). The role of dorsolateral prefrontal cortex in the selection of action as revealed by functional imaging. In S. Monsell & J. Driver (Eds.), *Control of Cognitive Processes* (pp. 549–565). Cambridge, MA: MIT Press.

Fucetola, R., Seidman, L. J., Kremen, W. S., Faraone, S. V., Goldstein, J. M., & Tsuang, M. T. (2000). Age and neuropsychologic function in schizophrenia: A decline in executive abilities beyond that observed in healthy volunteers. *Biological Psychiatry*, 48(2), 137–146.

Fuster, J. M. (2002). Physiology of executive functions: The perception-action cycle. In D. T. Stuss & R. T. Knight (Eds.), *Principles of Frontal Lobe Function* (pp. 96–108). New York: Oxford University Press.

Gansler, D. A., Varvaris, M., Swenson, L., & Schretlen, D. J. (2014). Cognitive estimation and its assessment. *Journal of Clinical and Experimental Neuropsychology*, 36(6), 559–568. doi:10.1080/13803395.2014.915933.

Gao, Y., & Raine, A. (2010). Successful and unsuccessful psychopaths: A neurobiological model. *Behavioral Sciences & the Law*, 28(2), 194–210.

Garavan, H., Ross, T., Murphy, K., Roche, R., & Stein, E. (2002). Dissociable executive functions in the dynamic control of behavior: Inhibition, error detection, and correction. *NeuroImage*, 17(4), 1820–1829.

Garon, N., Bryson, S. E., & Smith, I. M. (2008). Executive function in preschoolers: A review using an integrative framework. *Psychological Bulletin*, 134(1), 31–60. doi:10.1037/0033-2909.134.1.31.

Gathercole, S. E., Pickering, S. J., Ambridge, B., & Wearing, H. (2004). The structure of working memory from 4 to 15 years of age. *Developmental Psychology*, 40(2), 177–190.

Gilbert, S. J., & Burgess, P. W. (2008). Executive function. *Current Biology*, 18(3), R110–R114. doi:10.1016/j.cub.2007.12.014.

Glenn, A. L., Raine, A., & Schug, R. (2009). The neural correlates of moral decision-making in psychopathy. *Molecular Psychiatry*, 14, 5–6.

Goldberg, I. I., Harel, M., & Malach, R. (2006). When the brain loses its self: Prefrontal inactivation during sensorimotor processing. *Neuron*, 50(2), 329–339. doi: 10.1016/j.neuron.2006.03.015.

Goldman-Rakic, P. S. (1996). Regional and cellular fractionation of working memory. *Proceedings of the National Academy of Sciences of the United States of America*, 93(24), 13473–13480.

González-Blanch, C., Álvarez-Jiménez, M., Rodríguez-Sánchez, J. M., Pérez-Iglesias, R., Vázquez-Barquero, J. L., & Crespo-Facorro, B. (2006). Cognitive functioning in the early course of first-episode schizophrenia spectrum disorders. *European Archives of Psychiatry and Clinical Neuroscience*, 256(6), 364–371.

Greene, J., & Cohen, J. (2004). For the law, neuroscience changes nothing and everything. *Philosophical Transactions of the Royal Society of London. Series B, Biological Sciences*, 359(1451), 1775–1785.

Greicius, M. D., & Menon, V. (2004). Default-mode activity during a passive sensory task: Uncoupled from deactivation but impacting activation. *Journal of Cognitive Neuroscience*, 16(9), 1484–1492. doi:10.1162/0898929042568532.

Grosbras, M.-H., Jansen, M., Leonard, G., McIntosh, A., Osswald, K., Poulsen, C., et al. (2007). Neural mechanisms of resistance to peer influence in early adolescence. *Journal of Neuroscience*, 27(30), 8040–8045.

Guenther, L. (2012). The living death of solitary confinement. *New York Times Opinionator*. August 26, 2012. https://opinionator.blogs.nytimes.com/2012/08/26/the-living-death-of-solitary-confinement/

Guiney, H., & Machado, L. (2013). Benefits of regular aerobic exercise for executive functioning in healthy populations. *Psychonomic Bulletin & Review*, 20, 73–86.

Gusnard, D. A., Akbudak, E., Shulman, G. L., & Raichle, M. E. (2001). Medial prefrontal cortex and self-referential mental activity: Relation to a default mode of brain function. *Proceedings of the National Academy of Sciences of the United States of America*, 98(7), 4259–4264. doi:10.1073/pnas.071043098.

Gutowski, C., & Mills, S. (2014). Serial killer Brian Dugan gives 1st prison interview: "I could not stop." *Chicago Tribune*, December, 13, 2014. http://www.chicagotribune.com/news/ct-brian-dugan-serial-killer-interview-met-20141212-story.html.

Hagerty, B. B. (2010). Inside a psychopath's brain: The sentencing debate. *NPR Morning Edition*, June 30, 2010. https://www.npr.org/templates/story/story.php?storyId=128116806.

Haney, C. (2003). Mental health issues in long-term solitary and "supermax" confinement. *Crime and Delinquency*, 49(1), 124–156.

Happold, M. (2008). Child soldiers: Victims or perpetrators? *University of La Verne Law Review*, 29, 56–87.

Hare, R. D. (1965). Temporal gradient of fear arousal in psychopaths. *Journal of Abnormal Psychology*, 70(6), 442–445.

Hare, R. D. (1982). Psychopathy and physiological activity during anticipation of an aversive stimulus in a distraction paradigm. *Psychophysiology*, 19(3), 266–271.

Harlow, J. M. (1868). Recovery from the passage of an iron bar through the head. *Publications of the Massachusetts Medical Society*, 2, 327–346.

Hart, H. L. A. (1968). *Punishment and Responsibility: Essays in the Philosophy of Law*. New York: Oxford University Press.

Harvey, P. D., Koren, D., Reichenberg, A., & Bowie, C. R. (2006). Negative symptoms and cognitive deficits: What is the nature of their relationship? *Schizophrenia Bulletin*, 32(2), 250–258.

Hedden, T., & Gabrieli, J. D. (2004). Insights into the ageing mind: A view from cognitive neuroscience. *Nature Reviews. Neuroscience*, 5(2), 87–96.

Heinrichs, R. W., & Zakzanis, K. K. (1998). Neurocognitive deficit in schizophrenia: A quantitative review of the evidence. *Neuropsychology*, 12(3), 426–445.

Hermes, D., Miller, K. J., Wandell, B. A., & Winawer, J. (2014). Stimulus dependence of gamma oscillations in human visual cortex. *Cerebral Cortex*, 25(9), 2951–2959. doi:10.1093/cercor/bhu091.

Higginson, C. I., Thompson, T., Benjamin, A. E., Rosales, A. C., Zane, K. L., Valenti, M., et al. (2017). Construct validity of the Functional Assessment of Verbal Reasoning and Executive Strategies (FAVRES). *Brain Injury*, 31(13–14), 1807–1812. doi:10.1080/02699052.2017.1347278.

Hillman, C., Erickson, K., & Kramer, A. (2008). Be smart, exercise your heart: Exercise effects on brain and cognition. *Nature Reviews. Neuroscience*, 9, 58–65.

Hirstein, W. (2005). *Brain Fiction: Self-deception and the Riddle of Confabulation*. Cambridge, MA: MIT Press.

Hirstein, W. (2008). Mindmelding: Connected brains and the problem of consciousness. *Mens Sana Monographs*, 6(1), 110–130. doi:10.4103/0973-1229.38516.

Hirstein, W. (2011). The contribution of prefrontal executive processes to creating a sense of self. *Mens Sana Monographs*, 9(1), 150–158.

Hirstein, W. (2012). *Mindmelding: Consciousness, Neuroscience, and the Mind's Privacy*. Oxford: Oxford University Press.

Hirstein, W. (2015). Consciousness despite network underconnectivity in autism: Another case of consciousness without prefrontal activity? In R. Gennaro (Ed.), *Disturbed Consciousness: New Essays on Psychopathology and Theories of Consciousness* (pp. 249–263). Cambridge, MA: MIT Press.

Hirstein, W., & Sifferd, K. (2011). The legal self: Executive processes and legal theory. *Consciousness and Cognition*, 20(1), 156–171.

Hirstein, W., & Sifferd, K. (2014). Ethics and the brains of psychopaths: The significance of psychopathy for our ethical and legal theories. In C. Wolfe (Ed.), *Brain Theory: Essays in Critical Neurophilosophy* (pp. 149–170). New York: Palgrave-Macmillan.

Hirtenlehner, H., & Kunz, F. (2016). The interaction between self-control and morality in crime causation among older adults. *European Journal of Criminology*, 13(3), 393–409.

Hofmann, W., Schmeichel, B. J., & Baddeley, A. D. (2012). Executive functions and self-regulation. *Trends in Cognitive Sciences*, 16(3), 174–180. doi:10.1016/j.tics.2012.01.006.

Holtfreter, K., Reisig, M. D., & O'Neal, E. N. (2015). Prescription drug misuse in late adulthood: An empirical examination of competing explanations. *Journal of Drug Issues*, 45(4), 351–367.

Hoppenbrouwers, S. S., Bulten, B. H., & Brazil, I. A. (2016). Parsing fear: A reassessment of the evidence for fear deficits in psychopathy. *Psychological Bulletin*, 142(6), 573–600.

Hughes, V. (2010). Science in court: Head case. *Nature*, 464(7287), 340–342.

Huizinga, M., Dolan, C. V., & van der Molen, M. W. (2006). Age-related change in executive function: Developmental trends and a latent variable analysis. *Neuropsychologia*, 44(11), 2017–2036.

Hulme, O. J., Friston, K. F., & Zeki, S. (2009). Neural correlates of stimulus reportability. *Journal of Cognitive Neuroscience*, 21(8), 1602–1610.

Hume, D. (1739). *A Treatise of Human Nature, The Clarendon Edition of the Works of David Hume*. Oxford: Oxford University Press.

Hume, D. (1975). *Enquiries Concerning Human Understanding and Concerning the Principles of Morals*. Oxford: Clarendon Press. Original edition, 1777.

Hunt, K. S., & Dumville, R. (2016). *Recidivism among federal offenders: A comprehensive overview*. Washington, DC: United States Sentencing Commission.

Hutton, S., Puri, B., Duncan, L.-J., Robbins, T., Barnes, T., & Joyce, E. (1998). Executive function in first-episode schizophrenia. *Psychological Medicine*, 28(2), 463–473.

International Criminal Court. (2016). *Case Information Sheet: The Prosecutor v. Dominic Ongwen*. https://www.icc-cpi.int/uganda/ongwen/Documents/OngwenEng.pdf

Ireland, J. L., & Culpin, V. (2006). The relationship between sleeping problems and aggression, anger, and impulsivity in a population of young offenders. *Journal of Adolescent Health*, 38, 649–655.

Ishikawa, S. S., Raine, A., Lencz, T., Bihrle, S., & Lacasse, L. (2001). Autonomic stress reactivity and executive functions in successful and unsuccessful criminal psychopaths from the community. *Journal of Abnormal Psychology*, 110(3), 423–432.

Janssen, K. (2017). Cook County Jail inmates can order hot pizza delivered to their cells. *Chicago Tribune*, May 20, 2017. http://www.chicagotribune.com/news/chicagoinc/ct-pizza-in-jail-0510-chicago-inc-20170509-story.html.

Jennings, C. D. (2015). Consciousness without attention. *Journal of the American Philosophical Association*, 1(2), 276–295. doi:10.1017/apa.2014.14.

Johnson-Selfridge, M., & Zalewski, C. (2001). Moderator variables of executive functioning in schizophrenia: Meta-analytic findings. *Schizophrenia Bulletin*, 27(2), 305–316.

Jurjako, M., & Malatesti, L. (2016). Instrumental rationality in psychopathy: Implications from learning tasks. *Philosophical Psychology*, 29(5), 717–731.

Kane, M. J., & Engle, R. W. (2003). Working-memory capacity and the control of attention: The contributions of goal neglect, response competition, and task set to Stroop interference. *Journal of Experimental Psychology. General*, 132(1), 47–70.

Kennedy, A. M. (2003). Speech at the American Bar Association Annual Meeting. American Bar Association Annual Meeting, August 9, 2003.

King, M. (2009). The trouble with negligence. *Social Theory and Practice*, 35(4), 577–595.

Knausgaard, K. O. (2015). Anders Breivik's inexplicable crime. *The New Yorker*. May 25, 2015. https://www.newyorker.com/magazine/2015/05/25/the-inexplicable

Koch, C. (2004). *The Quest for Consciousness: A Neurobiological Approach*. Englewood, CO: Roberts and Company.

Koch, C., Massimini, M., Boly, M., & Tononi, G. (2016). Neural correlates of consciousness: Progress and problems. *Nature Reviews. Neuroscience*, 17(5), 307–321. doi:10.1038/nrn.2016.22.

Koch, C., & Tsuchiya, N. (2007). Attention and consciousness: Two distinct brain processes. *Trends in Cognitive Sciences*, 11(1), 16–22. doi:10.1016/j.tics.2006.10.012.

Koenigs, M., & Newman, J. P. (2013). The decision making impairment in psychopathy: Psychological and neurobiological mechanisms. In K. A. Kiehl & W. P. Sinnott-Armstrong (Eds.), *Handbook on Psychopathy and Law* (pp. 93–106). Oxford: Oxford University Press.

Koeske, Z. (2016). Cook County Jail chess program teaches inmates self-empowerment. *Chicago Tribune*, April 27, 2016. http://www.chicagotribune.com/suburbs/daily-southtown/news/ct-sta-jail-chess-st--0427-20160426-story.html.

Koziol, L. F., Barker, L. A., Joyce, A. W., & Hrin, S. (2014). Structure and function of large-scale brain systems. *Applied Neuropsychology. Child*, 3(4), 236–244. doi:10.1080/21622965.2014.946797.

Krishnadas, R., Ramanathan, S., Wong, E., Nayak, A., & Moore, B. (2014). Residual negative symptoms differentiate cognitive performance in clinically stable patients with schizophrenia and bipolar disorder. *Schizophrenia Research and Treatment.* doi:10.1155/2014/785310

Lahat, A. (2015). The neurocognitive development of moral judgments: The role of executive function. In J. Decety & T. Wheatley (Eds.), *The Moral Brain: A Multidisciplinary Perspective* (pp. 143–155). Cambridge, MA: MIT Press.

Laloyaux, J., Van der Linden, M., Levaux, M.-N., Mourad, H., Pirri, A., Bertrand, H., et al. (2014). Multitasking capacities in persons diagnosed with schizophrenia: A preliminary examination of their neurocognitive underpinnings and ability to predict real world functioning. *Psychiatry Research*, 217(3), 163–170.

Lamme, V. A. F. (2006). Towards a true neural stance on consciousness. *Trends in Cognitive Sciences*, 10(11), 494–501. doi:10.1016/j.tics.2006.09.001.

Langdon, R., & Coltheart, M. (2000). The cognitive neuropsychology of delusions. *Mind & Language*, 15(1), 184–218.

Larson, C. L., Baskin-Sommers, A. R., Stout, D. M., Balderston, N. L., Curtin, J. J., Schultz, D. H., et al. (2013). The interplay of attention and emotion: Top-down attention modulates amygdala activation in psychopathy. *Cognitive, Affective & Behavioral Neuroscience*, 13(4), 757–770.

Lee, J., & Park, S. (2005). Working memory impairments in schizophrenia: A meta-analysis. *Journal of Abnormal Psychology*, 114(4), 599–611.

Lee, M. Y. H. (2015). Yes, US locks people up at a higher rate than any other country. *The Washington Post*, July 7, 2015. https://www.washingtonpost.com/news/fact-checker/wp/2015/07/07/yes-u-s-locks-people-up-at-a-higher-rate-than-any-other-country/.

Leeson, V. C., Barnes, T. R., Harrison, M., Matheson, E., Harrison, I., Mutsatsa, S. H., et al. (2008). The relationship between IQ, memory, executive function, and processing speed in recent-onset psychosis: 1-year stability and clinical outcome. *Schizophrenia Bulletin*. Advance online publication.

Leimkuhler, M. E., & Mesulam, M. M. (1985). Reversible go-no go deficits in a case of frontal lobe tumor. *Annals of Neurology*, 18(5), 617–619. doi:10.1002/ana.410180518.

Levy, N. (2007). The responsibility of the psychopath revisited. *Philosophy, Psychiatry, & Psychology*, 14(2), 129–138.

Levy, N. (2011). Expressing who we are: Moral responsibility and awareness of our reasons for action. *Analytic Philosophy*, 52(4), 243–251.

Levy, N. (2014). *Consciousness and Moral Responsibility*. Oxford: Oxford University Press.

Levy, R., & Goldman-Rakic, P. S. (2000). Segregation of working memory functions within the dorsolateral prefrontal cortex. *Experimental Brain Research*, 133(1), 23–32. doi:10.1007/s002210000397.

Lezak, M. D. (1995). *Neuropsychological Assessment* (3rd ed.). Oxford: Oxford University Press.

Li, C.-S. R. (2004). Do schizophrenia patients make more perseverative than non-perseverative errors on the Wisconsin Card Sorting Test? A meta-analytic study. *Psychiatry Research*, 129(2), 179–190.

Libet, B. (1992). The neural time-factor in perception, volition and free will. *Revue de Metaphysique et de Morale*, 2, 255–272.

Libet, B. (1996). Neural processes in the production of conscious experience. In M. Velmans (Ed.), *The Science of Consciousness: Psychological, Neuropsychological, and Clinical Reviews* (pp. 96–117). London: Routledge.

Libet, B. (2009). *Mind Time: The Temporal Factor in Consciousness*. Cambridge, MA: Harvard University Press.

Lindquist, K. A., & Barrett, L. F. (2012). A functional architecture of the human brain: Emerging insights from the science of emotion. *Trends in Cognitive Sciences*, 16(11), 533–540. doi:10.1016/j.tics.2012.09.005.

Llinas, R., & Ribary, U. (1993). Coherent 40-Hz oscillation characterizes dream state in humans. *Proceedings of the National Academy of Sciences of the United States of America*, 90(5), 2078–2081. doi:10.1073/pnas.90.5.2078.

Llinás, R. R., Leznik, E., & Urbano, F. J. (2002). Temporal binding via cortical coincidence detection of specific and nonspecific thalamocortical inputs: A voltage-dependent dye-imaging study in mouse brain slices. *Proceedings of the National Academy of Sciences of the United States of America*, 99(1), 449–454. doi:10.1073/pnas.012604899.

Long, T. (2004). Moderate reasons-responsiveness, moral responsibility, and manipulation. In J. Keim-Campbell, M. O'Rourke, & D. Shier (Eds.), *Freedom and Determinism* (pp. 151–172). Cambridge, MA: MIT Press.

Lösel, F., & Schmucker, M. (2004). Psychopathy, risk taking, and attention: A differentiated test of the somatic marker hypothesis. *Journal of Abnormal Psychology*, 113(4), 522–529.

Lou, H. C., Nowak, M., & Kjaer, T. W. (2005). The mental self. In S. Laureys (Ed.), *The Boundaries of Consciousness: Neurobiology and Neuropathology* (pp. 197–204). Amsterdam: Elsevier.

Lovell, D., Johnson, L. C., & Cain, K. (2007). Recidivism of supermax prisoners in Washington state. *Crime and Delinquency*, *53*(4), 633–656.

Luciana, M. (2013). Adolescent brain development in normality and psychopathology. *Development and Psychopathology*, 25(4pt2), 1325–1345.

Luks, T. (2002). Evidence for anterior cingulate cortex involvement in monitoring preparatory attentional set. *NeuroImage*, 17(2), 792–802. doi:10.1016/s1053-8119(02)91210-3.

Luna, B., Padmanabhan, A., & O'Hearn, K. (2010). What has fMRI told us about the development of cognitive control through adolescence? *Brain and Cognition*, 72(1), 101–113. doi:10.1016/j.bandc.2009.08.005.

Magnin, M., Rey, M., Bastuji, H., Guillemant, P., Mauguiere, F., & Garcia-Larrea, L. (2010). Thalamic deactivation at sleep onset precedes that of the cerebral cortex in humans. *Proceedings of the National Academy of Sciences of the United States of America*, 107(8), 3829–3833. doi:10.1073/pnas.0909710107.

Maibom, H. L. (2008). The mad, the bad, and the psychopath. *Neuroethics*, 1(3), 167–184.

Manard, M., Bahri, M. A., Salmon, E., & Collette, F. (2016). Relationship between grey matter integrity and executive abilities in aging. *Brain Research*, 1642, 562–580.

Maquet, P. (2000). Functional neuroimaging of normal human sleep by positron emission tomography. *Journal of Sleep Research*, 9(3), 207–231. doi:10.1046/j.1365-2869.2000.00214.x.

Maquet, P., Ruby, P., Maudoux, A., Albouy, G., Sterpenich, V., Dang-Vu, T., et al. (2005). Human cognition during REM sleep and the activity profile within frontal and parietal cortices: A reappraisal of functional neuroimaging data. In S. Laureys (Ed.), *The Boundaries of Consciousness: Neurobiology and Neuropathology* (pp. 219–228). Amsterdam: Elsevier.

Markowitsch, H. J., & Kessler, J. (2000). Massive impairment in executive functions with partial preservation of other cognitive functions: The case of a young patient with severe degeneration of the prefrontal cortex. *Experimental Brain Research*, 133(1), 94–102. doi:10.1007/s002210000404.

Matar, E., & Lewis, S. J. G. (2017). REM sleep behaviour disorder: Not just a bad dream. *Medical Journal of Australia*, 207(6), 262–268. doi:10.5694/mja17.00321.

Matthews, D. (2015). Some of your farmers market food is grown by Cook County Jail inmates. *DNAinfo*, June 12, 2015. https://www.dnainfo.com/chicago/20150612/loop/cook-county-jail-inmates-sell-kale-knit-goods-at-downtown-farmers-market.

Matusow, H., Dickman, S. L., Rich, J. D., Fong, C., Dumont, D. M., Hardin, C., et al. (2013). Medication assisted treatment in US drug courts: Results from a nationwide survey of availability, barriers and attitudes. *Journal of Substance Abuse Treatment, 44*(5), 473–480.

McKay, R., Langdon, R., & Coltheart, M. (2007). Models of misbelief: Integrating motivational and deficit theories of delusions. *Consciousness and Cognition*, 16(4), 932–941.

McKay, T., & Nelson, M. (2010). Propositional Attitude Reports. Last modified March 21, 2014. https://plato.stanford.edu/entries/prop-attitude-reports/.

McTeague, L. M., Goodkind, M. S., & Etkin, A. (2016). Transdiagnostic impairment of cognitive control in mental illness. *Journal of Psychiatric Research*, 83, 37–46.

McTeague, L. M., Huemer, J., Carreon, D. M., Jiang, Y., Eickhoff, S. B., & Etkin, A. (2017). Identification of common neural circuit disruptions in cognitive control across psychiatric disorders. *American Journal of Psychiatry*, 174(7), 676–685.

Meijers, J., Harte, J., Jonker, F., & Meynen, G. (2015). Prison brain? Executive dysfunction in prisoners. *Frontiers in Psychology*, 6, 43.

Mele, A. R. (2006). *Free Will and Luck*. Oxford: Oxford University Press.

Melle, I. (2013). The Breivik case and what psychiatrists can learn from it. *World Psychiatry; Official Journal of the World Psychiatric Association (WPA)*, 12(1), 16–21.

Mendes, S. M. (2004). Certainty, severity, and their relative deterrent effects: Questioning the implications of the role of risk in criminal deterrence policy. *Policy Studies Journal: The Journal of the Policy Studies Organization*, 32(1), 59–74.

Menon, V., & Uddin, L. Q. (2010). Saliency, switching, attention and control: A network model of insula function. *Brain Structure & Function*, 214(5–6), 655–667. doi:10.1007/s00429-010-0262-0.

Milham, M. P., Banich, M. T., Webb, A., Barad, V., Cohen, N. J., Wszalek, T., et al. (2001). The relative involvement of anterior cingulate and prefrontal cortex in attentional control depends on nature of conflict. *Brain Research. Cognitive Brain Research*, 12(3), 467–473. doi:10.1016/s0926-6410(01)00076-3.

Miller, E. K. (2000). The prefrontal cortex and cognitive control. *Nature Reviews. Neuroscience*, 1(1), 59–65.

Miller, E. K., Freedman, D. J., & Wallis, J. D. (2002). The prefrontal cortex: Categories, concepts and cognition. *Philosophical Transactions of the Royal Society of London. Series B, Biological Sciences*, 357(1424), 1123–1136. doi:10.1098/rstb.2002.1099.

Miller, E. K., & Wallis, J. D. (2009). Executive function and higher-order cognition: Definition and neural substrates. In L. R. Squire (Ed.), *Encyclopedia of Neuroscience* (pp. 99–104). Oxford: Academic Press.

Miller, G. (2014). What it's like to spend 20 years listening to psychopaths for science. *Wired*, April 17, 2014.

Milner, B. (1982). Some cognitive effects of frontal-lobe lesions in man. *Philosophical Transactions of the Royal Society of London. Series B, Biological Sciences*, 298(1089), 211–226. doi:10.1098/rstb.1982.0083.

Minzenberg, M. J., Laird, A. R., Thelen, S., Carter, C. S., & Glahn, D. C. (2009). Meta-analysis of 41 functional neuroimaging studies of executive function in schizophrenia. *Archives of General Psychiatry*, 66(8), 811–822.

Mischel, W., Ebbesen, E. B., & Raskoff Zeiss, A. (1972). Cognitive and attentional mechanisms in delay of gratification. *Journal of Personality and Social Psychology*, 21(2), 204–218.

Miyake, A., & Friedman, N. P. (2012). The nature and organization of individual differences in executive functions: Four general conclusions. *Current Directions in Psychological Science*, 21(1), 8–14.

Miyake, A., Friedman, N. P., Emerson, M. J., Witzki, A. H., Howerter, A., & Wager, T. D. (2000). The unity and diversity of executive functions and their contributions to complex "frontal lobe" tasks: A latent variable analysis. *Cognitive Psychology*, 41(1), 49–100.

Monahan, K. C., Steinberg, L., & Cauffman, E. (2009). Affiliation with antisocial peers, susceptibility to peer influence, and antisocial behavior during the transition to adulthood. *Developmental Psychology*, 45(6), 1520–1530.

Monahan, K. C., Steinberg, L., & Piquero, A. R. (2015). Juvenile justice policy and practice: A developmental perspective. *Crime and Justice*, 44(1), 577–619.

Moore, M. S. (1997). *Placing Blame: A Theory of the Criminal Law*. Oxford: Oxford University Press.

Moore, M. S. (2014). The quest for a responsible responsibility test: Norwegian insanity law after Breivik. *Criminal Law and Philosophy*, 1–49. Advance online publication.

Morris, N. (1974). *The Future of Imprisonment*. Chicago: University of Chicago Press.

Morse, S. J. (2000). Deprivation and desert. In W. C. Heffernan & J. Kleinig (Eds.), *From Social Justice to Criminal Justice: Poverty and the Administration of Criminal Law* (pp. 114–160). Oxford: Oxford University Press.

Morse, S. J. (2002). Uncontrollable urges and irrational people. *Virginia Law Review*, 88(5), 1025–1078.

Morse, S. J. (2006a). Brain overclaim syndrome and criminal responsibility: A diagnostic note. *Ohio State Journal of Criminal Law*, 3, 397–412.

Morse, S. J. (2006b). Moral and legal responsibility and the new neuroscience. In J. Illes (Ed.), *Neuroethics: Defining the Issues in Theory, Practice, and Policy* (pp. 33–50). Oxford: Oxford University Press.

Morse, S. J. (2008a). Determinism and the death of folk psychology: Two challenges to responsibility from neuroscience. *Minnesota Journal of Law, Science & Technology*, 9(1), 1–36.

Morse, S. J. (2008b). Psychopathy and criminal responsibility. *Neuroethics*, 1(3), 205–212.

Morse, S. J. (2013a). Compatibilist criminal law. In T. A. Nadelhoffer (Ed.), *The Future of Punishment* (pp. 107–132). New York: Oxford University Press.

Morse, S. J. (2013b). A good enough reason: Addiction, agency and criminal responsibility. *Inquiry*, 56(5), 490–518.

Morse, S. J. (2015). Neuroscience, free will, and responsibility. In W. Glannon (Ed.), *Free Will and the Brain: Neuroscientific, Philosophical, and Legal Perspectives* (pp. 251–286). Cambridge: Cambridge University Press.

Morse, S. J. (2017). Law and the sciences of the brain/mind. In R. Brownsword, E. Scotford, & K. Yeung (Eds.), *The Oxford Handbook of Law, Regulation, and Technology* (pp. 153–176). Oxford: Oxford University Press.

Moscovitch, M., & Melo, B. (1997). Strategic retrieval and the frontal lobes: Evidence from confabulation and amnesia. *Neuropsychologia, 35*, 1017–1034.

Murray, D., & Nahmias, E. (2014). Explaining away incompatibilist intuitions. *Philosophy and Phenomenological Research*, 88(2), 434–467.

Muzur, A., Pace-Schott, E. F., & Hobson, J. A. (2002). The prefrontal cortex in sleep. *Trends in Cognitive Sciences*, 6(11), 475–481.

Nagin, D. S. (1998). Criminal deterrence research at the outset of the twenty-first century. *Crime and Justice*, 23, 1–42.

Nahmias, E. (2014). Is free will an illusion? Confronting challenges from the modern mind sciences. In W. Sinnott-Armstrong (Ed.), *Moral Psychology: Vol. 4. Freedom and Responsibility* (pp. 1–26). Cambridge, MA: MIT Press.

Nahmias, E., & Murray, D. (2010). Experimental philosophy on free will: An error theory for incompatibilist intuitions. In J. Aguilar, A. Buckareff, & K. Frankish (Eds.), *New Waves in Philosophy of Action* (pp. 189–215). New York: Palgrave-Macmillan.

National Conference of State Legislators. (2011). Principles of Effective State Sentencing and Corrections Policy: A report of the NCSL Sentencing and Corrections Work Group. National Conference of State Legislators. http://www.ncsl.org/documents/cj/pew/wgprinciplesreport.pdf.

Nelkin, D. K. (2016). Accountability and desert. *Journal of Ethics*, 20(1–3), 173–189.

Nichols, S., & Knobe, J. (2007). Moral responsibility and determinism: The cognitive science of folk intuitions. *Noûs*, 41(4), 663–685.

Niendam, T. A., Laird, A. R., Ray, K. L., Dean, Y. M., Glahn, D. C., & Carter, C. S. (2012). Meta-analytic evidence for a superordinate cognitive control network subserving diverse executive functions. *Cognitive, Affective & Behavioral Neuroscience*, 12(2), 241–268. doi:10.3758/s13415-011-0083-5.

Noë, A. (2010). *Out of Our Heads: Why You Are Not Your Brain, and Other Lessons from the Biology of Consciousness*. New York: Hill and Wang.

Ochoa, S., Haro, J. M., Huerta-Ramos, E., Cuevas-Esteban, J., Stephan-Otto, C., Usall, J., et al. (2014). Relation between jumping to conclusions and cognitive functioning in people with schizophrenia in contrast with healthy participants. *Schizophrenia Research*, 159(1), 211–217.

Orellana, G., & Slachevsky, A. (2013). Executive functioning in schizophrenia. *Frontiers in Psychiatry*, 4, 35. doi:10.3389/fpsyt.2013.00035.

Owen, A. M., Downes, J. J., Sahakian, B. J., Polkey, C. E., & Robbins, T. W. (1990). Planning and spatial working memory following frontal lobe lesions in man. *Neuropsychologia*, 28(10), 1021–1034. doi:10.1016/0028-3932(90)90137-d.

Owen, A. M., McMillan, K. M., Laird, A. R., & Bullmore, E. (2005). N-back working memory paradigm: A meta-analysis of normative functional neuroimaging studies. *Human Brain Mapping*, 25(1), 46–59. doi:10.1002/hbm.20131.

Pardo, M. S., & Patterson, D. (2013). *Minds, Brains, and Law: The Conceptual Foundations of Law and Neuroscience*. Oxford: Oxford University Press.

Parkin, A. J. (1984). Amnesic syndrome: A lesion-specific disorder? *Cortex*, 20(4), 479–508. doi:10.1016/s0010-9452(84)80053-2.

Pashler, H. (2000). Task-switching and multitask performance (tutorial). In S. Monsell & J. Driver (Eds.), *Control of Cognitive Processes* (pp. 277–308). Cambridge, MA: MIT Press.

Paus, T. (2005). Mapping brain maturation and cognitive development during adolescence. *Trends in Cognitive Sciences*, 9(2), 60–68.

Penfield, W. (1975). *Mystery of the Mind: A Critical Study of Consciousness and the Human Brain*. Princeton, NJ: Princeton University Press.

Penney, S. (2012). Impulse control and criminal responsibility: Lessons from neuroscience. *International Journal of Law and Psychiatry*, 35(2), 99–103.

Pennington, B. F., & Ozonoff, S. (1996). Executive functions and developmental psychopathology. *Journal of Child Psychology and Psychiatry, and Allied Disciplines*, 37(1), 51–87. doi:10.1111/j.1469-7610.1996.tb01380.x.

Pera-Guardiola, V., Batalla, I., Bosque, J., Kosson, D., Pifarré, J., Hernández-Ribas, R., et al. (2016). Modulatory effects of psychopathy on Wisconsin Card Sorting Test performance in male offenders with antisocial personality disorder. *Psychiatry Research*, 235, 43–48.

Pereboom, D. (2013). Free will skepticism and criminal punishment. In T. Nadelhoffer (Ed.), *The Future of Punishment* (pp. 49–78). New York: Oxford University Press.

Pereboom, D., & Caruso, G. (2018). Hard-incompatibilist existentialism: Neuroscience, punishment, and meaning in life. In G. Caruso & O. Flanagan (Eds.), *Neuroexistentialism: Meaning, Morals, and Purpose in the Age of Neuroscience* (pp. 193–222). Oxford: Oxford University Press.

Petrides, M. (2000). Dissociable roles of mid-dorsolateral prefrontal and anterior inferotemporal cortex in visual working memory. *Journal of Neuroscience*, 20(19), 7496–7503.

Petrides, M. (2005). Lateral prefrontal cortex: Architectonic and functional organization. *Philosophical Transactions of the Royal Society of London. Series B, Biological Sciences*, 360(1456), 781–795. doi:10.1098/rstb.2005.1631.

Pharo, H., Sim, C., Graham, M., Gross, J., & Hayne, H. (2011). Risky business: Executive function, personality, and reckless behavior during adolescence and emerging adulthood. *Behavioral Neuroscience*, 125(6), 970–978.

Pidd, H. (2012). Anders Behring Breivik describes Utøya massacre to Oslo court. *The Guardian*, April 20, 2012. https://www.theguardian.com/world/2012/apr/20/anders-behring-breivik-massacre-court.

Pillsbury, S. H. (2013). Talking about cruelty: The Eighth Amendment and juvenile offenders after Miller v. Alabama. *Loyola of Los Angeles Law Review, 46*, 885–932.

Pimentel, D. (2012). Criminal child neglect and the 'Free Range Kid': Is overprotective parenting the new standard of care? *Utah Law Review*, 2012, 947–999.

Piquero, A. R., Farrington, D. P., & Blumstein, A. (2007). *Key Issues in Criminal Career Research: New Analyses of the Cambridge Study in Delinquent Development*. Cambridge: Cambridge University Press.

Pitts, M. A., Metzler, S., & Hillyard, S. A. (2014). Isolating neural correlates of conscious perception from neural correlates of reporting one's perception. *Frontiers in Psychology*, 5, 1078. doi:10.3389/fpsyg.2014.01078.

Posner, M. I. (Ed.). (2011). *Cognitive Neuroscience of Attention*. New York: Guilford Press.

Posner, M. I., & Fan, J. (2008). Attention as an organ system. In J. R. Pomerantz (Ed.), *Topics in Integrative Neuroscience: From Cells to Cognition* (pp. 31–61). Cambridge: Cambridge University Press.

Prencipe, A., Kesek, A., Cohen, J., Lamm, C., Lewis, M. D., & Zelazo, P. D. (2011). Development of hot and cool executive function during the transition to adolescence. *Journal of Experimental Child Psychology*, 108(3), 621–637.

Quine, W. (1956). Quantifiers and propositional attitudes. *Journal of Philosophy, 53*(5), 177–187.

Radhu, N., Dominguez, L. G., Farzan, F., Richter, M. A., Semeralul, M. O., Chen, R., et al. (2015). Evidence for inhibitory deficits in the prefrontal cortex in schizophrenia. *Brain*, 138(2), 483–497.

Raichle, M. E., MacLeod, A. M., Snyder, A. Z., Powers, W. J., Gusnard, D. A., & Shulman, G. L. (2001). A default mode of brain function. *Proceedings of the National Academy of Sciences of the United States of America*, 98(2), 676–682. doi:10.1073/pnas.98.2.676.

Raine, A., Lencz, T., Bihrle, S., LaCasse, L., & Colletti, P. (2000). Reduced prefrontal gray matter volume and reduced autonomic activity in antisocial personality disorder. *Archives of General Psychiatry*, 57(2), 119–127.

Ravizza, S. M., & Carter, C. S. (2008). Shifting set about task switching: Behavioral and neural evidence for distinct forms of cognitive flexibility. *Neuropsychologia*, 46(12), 2924–2935. doi:10.1016/j.neuropsychologia.2008.06.006.

Raz, N., & Rodrigue, K. M. (2006). Differential aging of the brain: Patterns, cognitive correlates and modifiers. *Neuroscience and Biobehavioral Reviews*, 30(6), 730–748.

Redding, R. E. (2006). The brain-disordered defendant: Neuroscience and legal insanity in the twenty-first century. *American University Law Review*, 56, 51–127.

Reiss, A. J., Jr., & Farrington, D. P. (1991). Advancing knowledge about co-offending: Results from a prospective longitudinal survey of London males. *Journal of Criminal Law and Criminology*, 82, 360–395.

Rensink, R. A., O'Regan, J. K., & Clark, J. J. (2003). To see or not to see: The need for visual attention to perceive changes in scene. In B. J. Baars, W. P. Banks, & J. P. Newman (Eds.), *Essential Sources in the Scientific Study of Consciousness* (pp. 251–261). Cambridge, MA: MIT Press.

Richard, M. (1990). *Propositional Attitudes: An Essay on Thoughts and How We Ascribe Them*. Cambridge: Cambridge University Press.

Rizzolatti, G., Fadiga, L., Gallese, V., & Fogassi, L. (1996). Premotor cortex and the recognition of motor actions. *Brain Research. Cognitive Brain Research*, 3(2), 131–141. doi:10.1016/0926-6410(95)00038-0.

Rosen, D. M. (2009). Who is a child—The legal conundrum of child soldiers. *Connecticut Journal of International Law*, 25, 81–118.

Rosenthal, D. M. (1986). Two concepts of consciousness. *Philosophical Studies*, 49(3), 329–359. doi:10.1007/bf00355521.

Rosenthal, D. M. (2005). *Consciousness and Mind*. Oxford: Oxford University Press.

Roskies, A. (2006). Neuroscientific challenges to free will and responsibility. *Trends in Cognitive Sciences*, 10(9), 419–423.

Roskies, A. (2012). Don't panic: Self-authorship without obscure metaphysics. *Philosophical Perspectives*, 26(1), 323–342.

Rothbart, M. K., & Posner, M. I. (2001). Mechanism and variation in the development of attentional networks. In C. A. Nelson & M. Luciana (Eds.), *Handbook of Developmental Cognitive Neuroscience* (pp. 353–363). Cambridge, MA: MIT Press.

Rozas, A. X. P., Juncos-Rabadán, O., & González, M. S. R. (2008). Processing speed, inhibitory control, and working memory: Three important factors to account for age-related cognitive decline. *International Journal of Aging & Human Development*, 66(2), 115–130.

Ruff, H. A., & Rothbart, M. K. (2001). *Attention in Early Development: Themes and Variations*. Oxford: Oxford University Press.

Satpute, A. B., Shu, J., Weber, J., Roy, M., & Ochsner, K. N. (2012). The functional neural architecture of self-reports of affective experience. *Biological Psychiatry*, 73, 631–638.

Satterthwaite, T. D., Wolf, D. H., Erus, G., Ruparel, K., Elliott, M. A., Gennatas, E., et al. (2013). Functional maturation of the executive system during adolescence. *Journal of Neuroscience*, 33(41), 16249–16261.

Scepkowski, L. A., & Cronin-Golomb, A. (2003). The alien hand: Cases, categorizations, and anatomical correlates. *Behavioral and Cognitive Neuroscience Reviews*, 2(4), 261–277. doi:10.1177/1534582303260119.

Schenk, C. H., & Mahowald, M. W. (2002). REM sleep behavior disorder: Clinical, developmental, and neuroscience perspectives 16 years after its formal identification in *Sleep*. *Sleep*, 25(2), 120–138.

Scherf, K. S., Sweeney, J. A., & Luna, B. (2006). Brain basis of developmental change in visuospatial working memory. *Journal of Cognitive Neuroscience*, 18(7), 1045–1058.

Schwarz, J., & Beyette, B. (1997). *Brain Lock: Free Yourself from Obsessive-Compulsive Disorder*. New York: HarperCollins.

Scott, E. S., & Steinberg, L. D. (2009). *Rethinking Juvenile Justice*. Cambridge, MA: Harvard University Press.

Searle, J. R. (1992). *The Rediscovery of the Mind*. Cambridge, MA: MIT Press.

Seghier, M. L., Fagan, E., & Price, C. J. (2010). Functional subdivisions in the left angular gyrus where the semantic system meets and diverges from the default network. *Journal of Neuroscience*, 30(50), 16809–16817. doi:10.1523/jneurosci.3377-10.2010.

Selimbeyoglu, A., & Parvizi, J. (2010). Electrical stimulation of the human brain: Perceptual and behavioral phenomena reported in the old and new literature. *Frontiers in Human Neuroscience*, 4, 46. doi:10.3389/fnhum.2010.00046.

Semkovska, M., Bédard, M.-A., Godbout, L., Limoge, F., & Stip, E. (2004). Assessment of executive dysfunction during activities of daily living in schizophrenia. *Schizophrenia Research*, 69(2), 289–300.

Shackman, A. J., Salomons, T. V., Slagter, H. A., Fox, A. S., Winter, J. J., & Davidson, R. J. (2011). The integration of negative affect, pain and cognitive control in the cingulate cortex. *Nature Reviews. Neuroscience*, 12(3), 154–167.

Shallice, T. (1982). Specific impairments of planning. *Philosophical Transactions of the Royal Society of London. Series B, Biological Sciences*, 298(1089), 199–209.

Shallice, T. (1988). *From Neuropsychology to Mental Structure*. Cambridge: Cambridge University Press.

Shoemaker, D. (2015). *Responsibility from the Margins*. Oxford: Oxford University Press.

Siclari, F., Baird, B., Perogamvros, L., Bernardi, G., LaRocque, J. J., Riedner, B., et al. (2017). The neural correlates of dreaming. *Nature Neuroscience*, 20(6), 872–878. doi:10.1038/nn.4545.

Siclari, F., LaRocque, J. J., Postle, B. R., & Tononi, G. (2013). Assessing sleep consciousness within subjects using a serial awakening paradigm. *Frontiers in Psychology*, 4, 542. doi:10.3389/fpsyg.2013.00542.

Sifferd, K. L. (2006). In defense of the use of commonsense psychology in the criminal law. *Law and Philosophy*, 25, 571–612.

Sifferd, K. L. (2013). Translating scientific evidence into the language of the folk: Executive function as capacity-responsibility. In N. Vincent (Ed.), *Legal Responsibility and Neuroscience* (pp. 183–204). Oxford: Oxford University Press.

Sifferd, K. L. (2014). What does it mean to be a mechanism? Stephen Morse, nonreductivism, and mental causation. *Criminal Law and Philosophy*. Advance online publication.

Sifferd, K. L. (2016a). Unconscious mens rea: Lapses, negligence, and criminal responsibility. In D. Patterson & M. Pardo (Eds.), *Philosophical Foundations of Law and Neuroscience* (pp. 161–178). Oxford: Oxford University Press.

Sifferd, K. L. (2016b). Virtue ethics and criminal punishment. In J. Webber & A. Masala (Eds.), *From Personality to Virtue* (pp. 35–61). Oxford: Oxford University Press.

Sifferd, K. L., & Hirstein, W. (2013). On the criminal culpability of successful and unsuccessful psychopaths. *Neuroethics*, 6(1), 129–140.

Sifferd, K. L., Hirstein, W., & Fagan, T. K. (2017). Legal insanity and executive function. In M. D. White (Ed.), *The Insanity Defense: Multidisciplinary Views on Its History, Trends, and Controversies* (pp. 215–242). Santa Barbara, CA: Praeger.

Singer, W. (1993). Synchronization of cortical activity and its putative role in information processing and learning. *Annual Review of Physiology*, 55(1), 349–374.

Skeem, J., Johansson, P., Andershed, H., Kerr, M., & Louden, J. E. (2007). Two subtypes of psychopathic violent offenders that parallel primary and secondary variants. *Journal of Abnormal Psychology*, 116(2), 395–409.

Skeem, J. L., Polaschek, D. L., Patrick, C. J., & Lilienfeld, S. O. (2011). Psychopathic personality: Bridging the gap between scientific evidence and public policy. *Psychological Science in the Public Interest*, 12(3), 95–162.

Smith, A. (2005). Responsibility for attitudes: Activity and passivity in mental life. *Ethics*, 115(2), 236–271.

Smith, A. (2008). Control, responsibility, and moral assessment. *Philosophical Studies*, 138(3), 367–392.

Smith, E. E., & Jonides, J. (1999). Storage and executive processes in the frontal lobes. *Science*, 283, 1657–1661.

Spreng, R. N., & Schacter, D. L. (2011). Default network modulation and large-scale network interactivity in healthy young and old adults. *Cerebral Cortex*, 22(11), 2610–2621. doi:10.1093/cercor/bhr339.

Spreng, R. N., Sepulchre, J., Turner, G. R., Stevens, W. D., & Schacter, D. L. (2013). Intrinsic architecture underlying the relations among the default, dorsal attention, and frontoparietal control networks of the human brain. *Journal of Cognitive Neuroscience*, 25(1), 74–86.

Spreng, R. N., Stevens, W. D., Chamberlain, J. P., Gilmore, A. W., & Schacter, D. L. (2010). Default network activity, coupled with the frontoparietal control network, supports goal-directed cognition. *NeuroImage*, 53(1), 303–317. doi:10.1016/j.neuroimage.2010.06.016.

Sridharan, D., Levitin, D. J., & Menon, V. (2008). A critical role for the right fronto-insular cortex in switching between central-executive and default-mode networks. *Proceedings of the National Academy of Sciences of the United States of America*, 105(34), 12569–12574. doi:10.1073/pnas.0800005105.

Sripada, C. (2015). Moral responsibility, reasons, and the self. In D. Shoemaker (Ed.), *Oxford Studies in Agency and Responsibility* (pp. 242–264). Oxford: Oxford University Press.

Stace, W. T. (1953). The problem of free will. In *Religion and the Modern Mind* (pp. 248–258). London: Macmillan.

Starkstein, S. E., & Robinson, R. G. (1997). Mechanism of disinhibition after brain lesions. *Journal of Nervous and Mental Disease*, 185(2), 108–114. doi:10.1097/00005053-199702000-00007.

Steinberg, L. (2004). Risk taking in adolescence: What changes, and why? *Annals of the New York Academy of Sciences*, 1021(1), 51–58.

Steinberg, L. (2008). A social neuroscience perspective on adolescent risk-taking. *Developmental Review*, 28(1), 78–106.

Steinberg, L. (2010). A dual systems model of adolescent risk-taking. *Developmental Psychobiology*, 52(3), 216–224.

Steinberg, L. (2012). Should the science of adolescent brain development inform public policy? *Issues in Science and Technology*, 28(3), 67–78.

Steinberg, L., Albert, D., Cauffman, E., Banich, M., Graham, S., & Woolard, J. (2008). Age differences in sensation seeking and impulsivity as indexed by behavior and self-report: Evidence for a dual systems model. *Developmental Psychology*, 44(6), 1764–1778.

Steinberg, L., & Monahan, K. C. (2007). Age differences in resistance to peer influence. *Developmental Psychology*, 43(6), 1531–1543.

Steinberg, L., & Scott, E. S. (2003). Less guilty by reason of adolescence: Developmental immaturity, diminished responsibility, and the juvenile death penalty. *American Psychologist*, 58(12), 1009–1018.

Strawson, P. (1962). Freedom and resentment. *Proceedings of the British Academy*, 48, 1–25.

Stuss, D. T., & Alexander, M. P. (2007). Is there a dysexecutive syndrome? *Philosophical Transactions of the Royal Society of London. Series B, Biological Sciences*, 362(1481), 901–915. doi:10.1098/rstb.2007.2096.

Talbert, M. (2008). Blame and responsiveness to moral reasons: Are psychopaths blameworthy? *Pacific Philosophical Quarterly*, 89(4), 516–535.

Thomason, K. K. (2016). Guilt and child soldiers. *Ethical Theory and Moral Practice*, 19(1), 115–127.

Vaish, A., Carpenter, M., & Tomasello, M. (2016). The early emergence of guilt-motivated prosocial behavior. *Child Development*, 87(6), 1772–1782.

van Gaal, S., De Lange, F. P., & Cohen, M. X. (2012). The role of consciousness in cognitive control and decision making. *Frontiers in Human Neuroscience*, 6(121).

Vargas, M. (2013). *Building Better Beings: A Theory of Moral Responsibility*. Oxford: Oxford University Press.

Vera-Estay, E., Dooley, J., & Beauchamp, M. (2015). Cognitive underpinnings of moral reasoning in adolescence: The contribution of executive functions. *Journal of Moral Education*, 44(1), 17–33.

Vincent, J. L., Kahn, I., Snyder, A. Z., Raichle, M. E., & Buckner, R. L. (2008). Evidence for a frontoparietal control system revealed by intrinsic functional connectivity. *Journal of Neurophysiology*, 100(6), 3328–3342. doi:10.1152/jn.90355.2008.

Vincent, N. A. (2009). Responsibility: Distinguishing virtue from capacity. *Polish Journal of Philosophy*, 3(1), 111–126.

Vincent, N. A. (2011). A structured taxonomy of responsibility concepts. In N. A. Vincent, I. Van de Poel, & J. van den Hoven (Eds.), *Moral Responsibility* (pp. 15–35). Dordrecht: Springer.

Vincent, N. A. (2013). Blame, desert and compatibilist capacity: A diachronic account of moderateness in regards to reasons-responsiveness. *Philosophical Explorations*, 16(2), 1–17.

Vincent, N. A., van de Poel, I., & van den Hoven, J. (2011). *Moral Responsibility: Beyond Free Will and Determinism*. New York: Springer.

Wainryb, C. (2011). 'And so they ordered me to kill a person': Conceptualizing the impacts of child soldiering on the development of moral agency. *Human Development*, 54(5), 273–300.

Walker, N., & McCabe, S. (1968). *Crime and Insanity in England: The Historical Perspective*. Edinburgh: Edinburgh University Press.

Wallace, R. J. (1996). *Responsibility and the Moral Sentiments*. Cambridge: Harvard University Press.

Walsh, V., & Kulikowski, J. (Eds.). (1998). *Perceptual Constancy: Why Things Look as They Do*. Cambridge: Cambridge University Press.

Walmsley, R. (2015). *World Prison Population List* (11th ed.). Institute for Criminal Policy Research; http://www.prisonstudies.org/sites/default/files/resources/downloads/world_prison_population_list_11th_edition_0.pdf.

Watanabe, J., Sugiura, M., Sato, K., Sato, Y., Maeda, Y., Matsue, Y., et al. (2002). The human prefrontal and parietal association cortices are involved in NO-GO performances: An event-related fMRI study. *NeuroImage*, 17(3), 1207–1216. doi:10.1006/nimg.2002.1198.

Webber, J. (2006). Virtue, character and situation. *Journal of Moral Philosophy*, 3(2), 193–213.

Webster, T. (2007). Babes with arms: International law and child soldiers. *George Washington International Law Review*, 39, 227–254.

Wilson, B. A., Alderman, N., Burgess, P., Emslie, H., & Evans, J. J. (1996). *Behavioural Assessment of the Dysexecutive Syndrome*. St. Edmunds, UK: Thames Valley Test Company.

Wilson, B. A., Evans, J. J., Emslie, H., Alderman, N., & Burgess, P. (1998). The development of an ecologically valid test for assessing patients with a dysexecutive syndrome. *Neuropsychological Rehabilitation*, 8(3), 213–228. doi:10.1080/713755570.

Wolf, S. (1987). Sanity and the metaphysics of responsibility. In F. Schoeman (Ed.), *Responsibility, Character, and the Emotions* (pp. 46–62). Cambridge: Cambridge University Press.

Wolfe, S. E., Reisig, M. D., & Holtfreter, K. (2016). Low self-control and crime in late adulthood. *Research on Aging*, 38(7), 767–790.

Yang, Y., Raine, A., Colletti, P., Toga, A. W., & Narr, K. L. (2010). Morphological alterations in the prefrontal cortex and the amygdala in unsuccessful psychopaths. *Journal of Abnormal Psychology*, 119(3), 546–554

Yang, Y., Raine, A., Lencz, T., Bihrle, S., LaCasse, L., & Colletti, P. (2005). Volume reduction in prefrontal gray matter in unsuccessful criminal psychopaths. *Biological Psychiatry*, 57(10), 1103–1108.

Yeo, B. T., Krienen, F. M., Sepulcre, J., Sabuncu, M. R., Lashkari, D., Hollinshead, M., et al. (2011). The organization of the human cerebral cortex estimated by intrinsic functional connectivity. *Journal of Neurophysiology*, 106(3), 1125–1165. doi:10.1152/jn.00338.2011.

Young Adult Court, Superior Court of California—County of San Francisco. Accessed June 1, 2017. http://www.sfsuperiorcourt.org/divisions/collaborative/yac.

Young, A. W., & Block, N. (1996). Consciousness. In V. Bruce (Ed.), *Unsolved Mysteries of the Mind: Tutorial Essays in Cognition* (pp. 149–180). Hove, UK: Psychology Press.

Zahn-Waxler, C., & Robinson, J. (1995). Empathy and guilt: Early origins of feelings of responsibility. In J. P. Tangney & K. W. Fischer (Eds.), *Self-conscious Emotions: The Psychology of Shame, Guilt, Embarrassment, and Pride* (pp. 143–173). New York: Guilford Press.

Zane, S. N., Welsh, B. C., & Mears, D. P. (2016). Juvenile transfer and the specific deterrence hypothesis. *Criminology & Public Policy*, 15(3), 901–925.

Zeier, J. D., Baskin-Sommers, A. R., Hiatt Racer, K. D., & Newman, J. P. (2012). Cognitive control deficits associated with antisocial personality disorder and psychopathy. *Personality Disorders*, 3(3), 283–293.

Zeier, J. D., Maxwell, J. S., & Newman, J. P. (2009). Attention moderates the processing of inhibitory information in primary psychopathy. *Journal of Abnormal Psychology*, 118(3), 554–563.

Zeki, S., & Bartels, A. (1999). Toward a theory of visual consciousness. *Consciousness and Cognition*, 8(2), 225–259. doi:10.1006/ccog.1999.0390.

Zhan, C., Liu, Y., Wu, K., Gao, Y., & Li, X. (2017). Structural and functional abnormalities in children with attention-deficit/hyperactivity disorder: A focus on subgenual anterior cingulate cortex. *Brain Connectivity*, 7(2), 106–114. doi:10.1089/brain.2016.0444.

Zimring, F. E., & Laqueur, H. (2015). Kids, groups, and crime: In defense of conventional wisdom. *Journal of Research in Crime and Delinquency*, 52(3), 403–413.

Name Index

Subject Index